Voltage-Enhanced Processing of Biomass and Biochar

Wiley-ASME Press Series

Corrosion and Materials in Hydrocarbon Production: A Compendium of Operational and Engineering Aspects
Bijan Kermani, Don Harrop

Design and Analysis of Centrifugal Compressors
Rene Van den Braembussche

Case Studies in Fluid Mechanics with Sensitivities to Governing Variables
M. Kemal Atesmen

The Monte Carlo Ray-Trace Method in Radiation Heat Transfer and Applied Optics
J. Robert Mahan

Dynamics of Particles and Rigid Bodies: A Self-Learning Approach
Mohammed F. Daqaq

Primer on Engineering Standards, Expanded Textbook Edition
Maan H. Jawad, Owen R. Greulich

Engineering Optimization: Applications, Methods and Analysis
R. Russell Rhinehart

Compact Heat Exchangers: Analysis, Design and Optimization using FEM and CFD Approach
C. Ranganayakulu, Kankanhalli N. Seetharamu

Robust Adaptive Control for Fractional-Order Systems with Disturbance and Saturation
Mou Chen, Shuyi Shao, Peng Shi

Robot Manipulator Redundancy Resolution
Yunong Zhang, Long Jin

Stress in ASME Pressure Vessels, Boilers, and Nuclear Components
Maan H. Jawad

Combined Cooling, Heating, and Power Systems: Modeling, Optimization, and Operation
Yang Shi, Mingxi Liu, Fang Fang

Applications of Mathematical Heat Transfer and Fluid Flow Models in Engineering and Medicine
Abram S. Dorfman

Bioprocessing Piping and Equipment Design: A Companion Guide for the ASME BPE Standard
William M. (Bill) Huitt

Nonlinear Regression Modeling for Engineering Applications: Modeling, Model Validation, and Enabling Design of Experiments
R. Russell Rhinehart

Geothermal Heat Pump and Heat Engine Systems: Theory and Practice
Andrew D. Chiasson

Fundamentals of Mechanical Vibrations
Liang-Wu Cai

Introduction to Dynamics and Control in Mechanical Engineering Systems
Cho W.S. To

Voltage-Enhanced Processing of Biomass and Biochar

Gerardo Diaz
University of California
Merced, California
USA

This Work is a co-publication between ASME Press and John Wiley & Sons Ltd

Registered Office
John Wiley & Sons, Inc., 111 River Street, Hoboken, NJ 07030, USA
John Wiley & Sons Ltd, The Atrium, Southern Gate, Chichester, West Sussex, PO19 8SQ, UK

Editorial Office
111 River Street, Hoboken, NJ 07030, USA

For details of our global editorial offices, customer services, and more information about Wiley products visit us at www.wiley.com.

Wiley also publishes its books in a variety of electronic formats and by print-on-demand. Some content that appears in standard print versions of this book may not be available in other formats.

Library of Congress Cataloging-in-Publication Data applied for
Hardback ISBN: 9781119739739

Cover Design: Wiley
Cover Image: Courtesy of Gerardo Diaz

Set in 9.5/12.5pt STIXTwoText by Straive, Chennai, India
Printed and bound by CPI Group (UK) Ltd, Croydon, CR0 4YY

C9781119739739_280622

To my wife, Kathleen, and my two wonderful children

Contents

Preface

One of the biggest lessons learned from the COVID-19 pandemic is that it has highlighted how vulnerable humanity is with respect to potential threats that had been predicted for decades and that we thought we were prepared to resist. Another important lesson is that shelter-in-place requirements mandated by governments around the world showed the impact of human activity in air quality and carbon emissions to the atmosphere. Clean skies were seen in places where a thick layer of smog was a common daily sight. The recent sixth assessment report from the Intergovernmental Panel on Climate Change concluded that widespread and rapid changes have occurred unequivocally due to human influence in warming the atmosphere, ocean, and land. As long as world leaders do not take strong action to limit carbon emissions to the atmosphere, we will continue to live in a world threatened by climate change, which will end up exposing more vulnerabilities of our society. Just in the United States, it is estimated that around 1 billion dry tones of biomass per year could be produced sustainably. This is in addition to the already available biomass that decomposes releasing methane and other pollutants to the atmosphere. The conversion of biomass to useful forms of energy such as electricity and heat, as well as the production of value-added products such as biochar and activated carbon, constitute a viable way to reduce biomass, generate renewable energy, and sequester carbon in a stable form. This book provides an overview of conventional biomass processing techniques as well as a description of technologies that utilize voltages and currents to enhance processing capabilities. The term plasma processing of biomass is usually associated with thermal plasma torches used for gasification of organic material. This book not only describes thermal plasma processing of biomass, but it also presents applications where nonthermal plasma discharges can be utilized in biomass processing plants, and applications where Joule heating of carbonaceous materials can be implemented. The book is intended for senior level undergraduate students and first year graduate students, who might not have a background in plasma, but are familiar with concepts of calculus, differential equations, and numerical algorithms. Chapter 1 provides a description of relevant properties of biomass, biochar, and activated carbon, while Chapter 2 gives a description of conventional methods of processing biomass and biochar. Chapter 3 provides an introduction to plasmas for thermal and nonthermal discharges, and Chapter 4 describes technologies that are suitable for utilizing the effects of applied voltages to enhance biomass processing. As properties of biomass vary after thermochemical decomposition, yielding a material with better electrical properties, Chapter 5 focuses on the analysis of the effects of applying voltages in processing of biochar. Thermal runaway behavior can be obtained with heating rates not achievable by conventional heating techniques. Chapter 6 provides an introduction of numerical simulation of plasmas. Finally, the inherit variability and even chaotic behavior of thermal arcs are analyzed in Chapter 7 in the context of the development of control techniques that can stabilize these discharges.

Gerardo Diaz

January 20, 2022　　　　　　　　　　Merced, California

Acknowledgments

I would like to acknowledge the following researchers as well as former and current graduate students who have collaborated with me in the past. Dr. Edbertho Leal-Quiros, Dr. YangQuan Chen, Dr. Carlos F.M. Coimbra, Dr. Williams R. Calderón-Muñoz, Todd Foret, Dr. Hugh McLaughlin, Greg Stangl, Dr. Sergio M. Pineda, Dr. Neeraj Sharma, Dr. Andres Munoz-Hernandez, Dr. Viacheslav Plotnikov, Dr. Sai Kiran Hota, Hector Gomez, Ziad Nasef, Kai Oh, Dr. Nararianto Indrawan, and Dr. Ajay Kumar. In addition, I would like to acknowledge the involvement of a large number of motivated undergraduate students who have helped with some of the experiments over the past years.

Acronyms

AC	alternating current
AI	artificial intelligence
APGD	atmospheric pressure glow discharge
BET	Brunauer, Emmett, and Teller
CEC	cation exchange capacity
DBD	dielectric barrier discharge
DC	direct current
DR	Dubinin–Radushkevich
EHD	electrohydrodynamics
GAC	granular activated carbon
HHV	high heating value
ID	inside diameter
IMC	internal model control
LHV	low heating value
LTE	local thermodynamic equilibrium
MOSFET	metal oxide semiconductor field effect transistor
MSW	municipal solid waste
NTP	nonthermal plasma
OD	outside diameter
PID	proportional-integral-differential
RF	radio frequency
TC	thermocouple
TLUD	Top-Lit UpDraft
UV	ultraviolet light
VOC	volatile organic compound
ZVS	zero voltage switching

Introduction

It has become evident that over the past decades, the impacts of climate change are increasing in severity and frequency. This situation has surpassed the threshold of affecting just our comfort level with higher average temperatures, heat waves, and modified precipitation patterns, but it is starting to threaten our livelihood. Climate change has been considered by some researchers as the biggest environmental challenge of our existence. For instance, average ambient temperatures continue to increase, severe drought conditions are occurring in several areas of the world, and six out of ten of the most extreme historical floods have taken place in the last 25 years. In addition, wildfire frequency and intensity are also increasing, partly due to climate change, but also due to outdated forest management practices and a large supply of biomass. It is here where there is great potential to utilize carbon-negative processes to reduce emissions and sequester carbon in a stable way. Well-established biomass processing techniques include combustion, gasification, pyrolysis, hydrolysis and hydrothermal liquefaction, which are suitable for a variety of applications that require steam, process heat, electricity, biochar, fuel gases, or synthetic liquid fuels. However, the utilization of voltage-driven techniques for the processing of biomass and biochar has been shown to have advantages for certain applications. This book concentrates on voltage-enhanced processing of carbonaceous materials, describing aspects related to thermal and non-thermal plasmas as well as the effects of Joule heating in the temperature distribution and conversion rate. In certain cases, it is necessary that the plasma discharge provides most of the energy required for the conversion. For these cases, a brief description of thermal plasma torches available is provided and experimental results of the conversion utilizing steam plasma are described. Results are compared against a thermodynamic model that predicts synthesis gas composition under the presence of a thermal plasma discharge. Simulation results of Joule heating of biomass, biochar and pyrolytic graphite are also provided. The thermochemical conversion of carbonaceous materials can also be enhanced with nonthermal plasma (NTP), in which the presence of the discharge generates ionized and excited species, radicals, etc., that are not present in conventional conversion processes. The purpose of the plasma in this case is not to provide the entire energy for the process but to enhance conversion. The book provides a description of the way that voltage is used to generate a NTP discharge, which exhibits highly energetic electrons with ions and neutral species at near-ambient temperature. A description of the physics related to these discharges is provided with experimental and simulation results of biomass gasification and plasma activation. Results related to tar breakdown are also provided as NTP is used to reduce pollutant emissions and to increase the fraction of hydrogen in synthesis gas by decomposing tars. An introduction to numerical simulations of non-equilibrium plasma discharges is provided and, a brief description of the control of these discharges is provided in the last chapter.

1

Carbonaceous Material Characterization

1.1 Material Characterization

The thermochemical conversion of biomass by means of processes such as torrefaction, pyrolysis, or gasification, produces a char-like material with properties that differ considerably from the original feedstock. Further processing in the form of chemical or physical activation continues to modify the properties of the carbonaceous materials produced. The following Sections (1.1.1–1.1.4) provide a description of the main properties needed to characterize carbon-based materials for applications of biochar production and energy conversion.

1.1.1 Thermophysical Properties

Thermophysical properties are directly related to the structure and composition of the carbon-based materials. There is a large volume of studies that have analyzed their variation due to the effects of thermochemical conversion processes (Balogun et al., 2018). These properties not only have an impact in the operating conditions of processing equipment but also affect transportation costs and pollutant emissions. The thermal and physical properties have a strong dependence on parameters such as moisture content and temperature (James, 1975; Skaar, 1988; Dietenberger et al., 1999; Zelinka et al., 2007); therefore, various models have been developed to represent them as a function of these factors. The most relevant properties include thermal conductivity, density, and specific heat. However, a more detailed description of properties for biomass can be found in Ross (2010), De Jong (2014), Goss and Miller (1992), and Gaur and Reed (1995).

Voltage-Enhanced Processing of Biomass and Biochar, First Edition. Gerardo Diaz.
© 2022 John Wiley & Sons Ltd. Published 2022 by John Wiley & Sons Ltd.

1.1.2 Moisture Content

The moisture content is defined as the mass of moisture (water) divided by the ovendry mass of the sample (Goss and Miller, 1992). The moisture content can be obtained by placing the sample in an oven filled with inert gas at a temperature of 105 °C for 24 hours, and then applying Eq. (1.1) to calculate the fraction of moisture in the sample.

$$MC = \frac{m_{wet} - m_{dry}}{m_{dry}} \times 100 \ (\%) \tag{1.1}$$

where m_{wet} is the mass of the specimen at a given moisture content and m_{dry} is the mass of the ovendry specimen.

1.1.3 Ultimate and Proximate Analysis

The description of the composition of carbonaceous materials is usually obtained by performing an *ultimate* and *proximate* analysis. The ultimate analysis provides the chemical composition of a material, providing information about the contents of carbon, hydrogen, nitrogen, oxygen (by difference), and sulfur of a dry sample on a weight basis (CHNOS analysis). In addition, the analysis provides the heating value of the sample, which provides the energy released due to combustion of the material. The high heating value (HHV) considers that water present in the sample as well as water formed during the combustion process are in liquid state. On the other hand, the low heating value (LHV) considers that water has not been condensed. The proximate analysis provides the information about moisture, volatile matter, fixed carbon, and ash content of a particular sample. A detailed description of these analyses can be found in De Jong (2014).

1.1.4 Dielectric and Electrical Properties

Dielectric properties are important for the study of storage and dissipation of electric energy in materials (Bain and Chand, 2017). A dielectric is an insulating material that is a very poor conductor of electric current. Wood and other types of biomass behave as dielectrics, but as high voltages are applied, breakdown can occur, which constitutes an irreversible change of the material that allows current to flow through it. The main dielectric properties for applications of thermochemical conversion are the relative permittivity and loss tangent. The relative permittivity (also called dielectric constant) describes the ability of a material to absorb and store energy from an applied electric field (James, 1975), and it is defined as (Barker-Jarvis et al., 2001):

$$\epsilon_r = \frac{\epsilon}{\epsilon_0}$$
$$= \epsilon_r' - j\epsilon_r'' \tag{1.2}$$

where ϵ_r is the relative permittivity, ϵ is the absolute permittivity of the material, $\epsilon_0 = 8.854 \times 10^{-12}$ (F/m) is the permittivity of vacuum, and ϵ_r' and ϵ_r'' are the real and imaginary parts of the relative permittivity. The relative permittivity can be represented by a scalar for a perfectly insulating material subject to DC voltage, but for AC voltage it is represented as a complex number given by Eq. (1.2), that varies significantly with respect to the applied frequency. The loss tangent denotes the dissipation of electric energy of a material and is defined as:

$$\tan(\delta) = \frac{\epsilon_r''}{\epsilon_r'} \tag{1.3}$$

In addition, as biomass is subject to thermochemical conversion, the char-like material produced has a higher electrical conductivity (an electrical property) than the original feedstock. When an electric field is applied to a material and a flow of current is established, part of the energy is dissipated as heat. As the resistivity of the material decreases due to thermal decomposition, the electrical conductivity characterizes the ability of the material to conduct electricity.

1.2 Biomass

Biomass is composed of renewable organic material that comes from plants and animals.[1] The most common types of biomass used for power or heat generation come from forest or agricultural waste. Municipal solid waste (MSW) is not considered in this chapter due to the mixing of organic materials with other types of waste. The description of the dependency of biomass properties with respect to temperature and moisture is described in this section.

Thermal conductivity: The thermal conductivity corresponds to the intrinsic ability of a material to conduct heat. According to the *Handbook of Wood* (Ross, 2010), for wood with moisture contents below 25%, the thermal conductivity can be obtained with the expression:

$$k = G_x(B + Cx) + A \text{ (W/(m K))} \tag{1.4}$$

---

1 https://www.eia.gov/energyexplained/biomass/.

where G_x is the specific gravity at moisture content x (in %), and where $A = 0.018\,64$, $B = 0.1941$, and $C = 0.004\,064$, for $G_x > 0.3$, and a temperature around $24\,°C$, with $x < 25\%$.

Specific heat: The specific heat can be interpreted as the amount of heat required per unit of mass to raise the temperature of a sample by one degree (Sonntag et al., 1998). Its value as a function of temperature for *dry wood* can be approximated by the expression:

$$c_{p_{dry}} = 0.0131 + 0.003\,867T \ (kJ/(kg\ K)) \tag{1.5}$$

where the temperature T is in Kelvin. For *wood* that contains a percentage x of moisture, the specific heat can be calculated as:

$$c_{p_x} = \frac{c_{p_{dry}} + c_{p_w}\frac{x}{100}}{1 + \frac{x}{100}} + A_c \tag{1.6}$$

where $c_{p_w} = 4.18\,kJ/(kg\ K)$ is the specific heat of water and A_c is a correction factor equal to:

$$A_c = x(b_1 + b_2T + b_3x) \tag{1.7}$$

where $b_1 = -0.061\,91$, $b_2 = 2.36 \times 10^{-4}$ and $b_3 = -1.33 \times 10^{-4}$.

Electrical conductivity: The electrical conductivity of wood depends on temperature and it can be approximated as (Skaar, 1988):

$$\sigma(T_L) = 371\,535.2 \times 10^{-3660/T_L} \tag{1.8}$$

where σ has units of $(\Omega\ m)^{-1}$. The dielectric constant can be estimated by a constant value of 800 (James, 1975; Dietenberger et al., 1999).

Tables 1.1–1.4 provide a list of properties for a number of biomass types.

1.3 Biochar

When biomass is heated to temperatures in the range between $200 < T < 800\,°C$ in conditions of limited or no oxygen supply, char formation exists (White and Dietenberger, 2001), and its properties change dramatically with respect to the original biomass feedstock. Biochar is a material produced with the intention of using it as soil amendment. However, in many published works, this material is still referred to as biochar even if it is produced for other purposes such as liquid or gas filtration or as construction material.

In this chapter, average properties of biochar produced at $T_{carb} = 800\,°C$ are summarized in Table 1.1. A more detailed summary of biochar properties can be found (Yang et al., 2017).

Table 1.1 List of thermophysical properties

Material	Specific gravity	k W/(m K) (12% MC)	Specific heat[a] kJ/(kg K)
Hardwoods[b]			1.63
Ash (white)	0.63	0.17	
Birch (yellow)	0.66	0.18	
Elm (rock)	0.67	0.18	
Maple (sugar)	0.66	0.18	
Oak (red)	0.65	0.18	
Oak (white)	0.72	0.19	
Softwoods[b]			1.63
Cedar (Western red)	0.33	0.10	
Douglas-fir (coast)	0.51	0.14	
Pine (ponderosa)	0.42	0.12	
Reedwood (old growth)	0.41	0.12	
Biochar (peach pits)	0.5	0.38	1.1
Graphite	1.78	113	0.72

a) Goss and Miller (1992).
b) Ross (2010).

Table 1.2 List of dielectric and electrical properties

Material	ϵ_r	$\tan(\delta)$	σ $(\Omega\ m)^{-1}$
Douglas-fir[a]	4.3 at 20 Hz	0.033 at 20 Hz	
Douglas-fir[a]	3.8 at 1 kHz	0.024 at 1 kHz	
Douglas-fir[b]	1.9 at 1 MHz	0.023 at 1 MHz	
Oak[a]	4.1 at 20 Hz	0.028 at 20 Hz	
Oak[a]	3.5 at 1 kHz	0.028 at 1 kHz	
Oak[a]	3.4 at 1 GHz	0.076 at 1 MHz	
Wood (damp)[c]			10^{-4} to 10^{-3}
Wood (ovendry)[c]			10^{-16} to 10^{-14}
Biochar[d]	42		2.6×10^1
Carbon (graphite)[c]	10–15 at 1 kHz		2 to 3×10^5

a) Goss and Miller (1992).
b) https://www.rfcafe.com/references/electrical/dielectric-constants-strengths.htm.
c) https://www.thoughtco.com/table-of-electrical-resistivity-conductivity-608499.
d) Win and Nang (2017).

Table 1.3 Ultimate analysis (weight fractions) and heating value of biomass analyzed in this work

Material	C (%)	H (%)	O (%)	N (%)	LHV (kJ/kg)
Biomass[a]					
Hard wood shaving	48.41	6.28	41.1	0.13	18 854.5
Peach pits	52.52	6.18	39.74	0.38	20 754.9
Almond hulls	44.31	5.64	40.13	1.06	18 040.5
Grape pomace	52.74	6.23	33.48	2.14	21 883.0
Coffee ground	56.13	7.16	27.03	2.53	23 771.7
Biochar					
Ponderosa pine	77.73	2.43	9.5	0.55	—
Walnut shell	78.86	3.48	13.7	0.64	—
Peach pits[b]	65.62	2.33	10.2	0.57	—
Almond shell[c]	50.5	6.6	—	0.21	18 200 (HHV)
Almond pruning[c]	51.3	5.7	—	0.77	18 200 (HHV)
Activated carbon					
Paulownia wood[d]	70.83	3.41	25.76	—	23 300
Peach pits[b]	63.17	0.10	1.44	0.32	—

a) Diaz et al. (2015).
b) Munoz-Hernandez (2018).
c) Chen et al. (2010).
d) Yorgun and Yildiz (2015).

Thermal conductivity: The thermal conductivity of biochar depends on temperature approximately as follows (Ragland et al., 1991; Parfen'eva et al., 2011):

$$k_L(T_L) = 0.0013T_L - 0.01 \ (\text{W}/(\text{m K})) \tag{1.9}$$

Specific heat: The specific heat of biochar as a function of temperature can be found in Dupont et al. (2014).

Electrical conductivity: It has been reported that the electrical conductivity of wood has a strong dependence on temperature (Dietenberger et al., 1999). However, the electrical conductivity of dry wood is so low that it still remains a good electrical insulator at moderate temperatures. Nonetheless, when wood is heated, it becomes charcoal and the electrical conductivity increases by several orders of magnitude (Kwon et al., 2013; Parfen'eva et al., 2011; Kumar and Gupta, 1993). For instance, measured

Table 1.4 Proximate analysis of biomass analyzed (weight fractions)

Material	Ash (%)	Fixed carbon (%)	Moisture (%)	Sulfur (%)	Volatile matter (%)
Biomass[a]					
Hard wood shaving	4.03	7.53	5.72	<0.01	7.53
Peach pits	0.72	11.9	36.73	0.46	50.7
Almond hulls	8.86	18.9	8.01	<0.01	64.3
Grape pomace	5.29	18.5	8.97	0.12	67.2
Coffee ground	1.14	7.07	54.69	0.08	37.1
Almond shell[b]	0.6	15.8	3.3	—	80.3
Almond pruning[b]	1.2	15.9	10.6	—	72.2
Biochar					
Peach pits[c]	21.24	54.98	Dry basis	—	23.78
Activated carbon					
Paulownia wood[d]	2.63	77.27	2.30	—	17.80
Peach pits[c]	35.0	59.96	Dry basis	—	5.04

a) Diaz et al. (2015).
b) Chen et al. (2010).
c) Munoz-Hernandez (2018).
d) Yorgun and Yildiz (2015).

at room temperature, the electrical conductivity of oven dry wood is of the order of 10^{-15} $(\Omega\ \text{m})^{-1}$ (Dietenberger et al., 1999), whereas the electrical conductivity of biochar produced at $T_{carb} = 600\ ^\circ\text{C}$ is roughly 7×10^{-4} $(\Omega\ \text{m})^{-1}$ (Kwon et al., 2013). The electrical conductivity of biochar depends on temperature as (Sugimoto and Norimoto, 2004; Kwon et al., 2013):

$$\sigma(T_L) = 64\ 565 \times 10^{-1000/T_L} \tag{1.10}$$

where the units of σ are $(\Omega\ \text{m})^{-1}$.

Tables 1.1–1.4 provide a list of properties for a number of biochar types.

1.3.1 Surface Area, Cation Exchange Capacity, and pH

As mentioned above, biochar is a material produced with the intention of using it as soil amendment. The ability of biochar to increase nutrient retention for plants is related to the *cation exchange capacity* (CEC) property. In addition, as biomass is thermally decomposed, a higher fraction of hydrogen and oxygen are removed from the material as compared to carbon. Due to this, the overall carbon concentration increases and it is referred to as *carbon recovery*. From initial carbon fractions around 39%, it can be increased to values in the range between 60% and 85%.

Another important aspect is that low-pH organic matter in soil shows very low values of CEC (Lehmann, 2007). Therefore, biochar pH is another important property that can improve soil health. Finally, climate change increases the odds of worsening drought conditions in several parts of the world. Therefore, a material with adequate surface area can help to increase water-holding capacity. Depending on how it was produced, biochar can have a relatively large surface area that can potentially help in reducing the amount of irrigation water used for plants. One of the most common methods used to determine surface area of biochar is by determining its BET *surface area* utilizing the equation and methodology developed by Brunauer, Emmett, and Teller that utilizes N_2 at 77 K, where results are provided in m^2/g of biochar. It is important to indicate that the surface area determined is related to the adsorption of the nitrogen molecule in the material, and thus, depends on the adsorbate utilized. For nitrogen at 77 K, the activation energy of adsorption is very high (greater than 40 kJ/mol) and thus there is restricted adsorption into narrow porous (diameter, $d < 0.5$ nm) requiring very long equilibrium times (weeks to years) (Marsh and Reinoso, 2006). Therefore, many studies use adsorption of carbon dioxide at 298 K based on the equation by Dubinin–Radushkevich (DR) to determine surface area (m^2/g) of highly porous biochar and activated carbon. These materials with high levels of microporosity are usually utilized for applications that require removing pollutants from liquid or gas streams.

There are other methods that refer to the ability of porous material, such as biochar or activated carbon, to adsorb a gas. For instance the ASTM D5742 standard is used to determine *butane activity* and provides the ratio (in %) between the mass of butane adsorbed and the mass of the original sample of biochar or activated carbon.

Very common in the industrial sector, the *iodine number* measures the micropore content of biochar or activated carbon (0–20 Å, or up to 2 nm) by adsorption of iodine, where results are often reported in mg/g (Mianowski et al., 2007).

Finally, adsorption of R134a at ambient temperature has also been proposed to characterize carbonaceous materials, where results are provided in percentage of weight gained.

Other ways of characterizing biochar include moisture content, ash, mobile matter, and resident matter (McLaughlin, 2010), in addition to physical attributes such as weight, hardness, greasiness to the touch, and morphological descriptions such as size, characteristic length, and shape.

1.4 Activated Carbon

For applications related to pollutant absorption from liquid or gas streams, a material with very high surface area and microporosity (porous diameter, $d < 2\,nm$) is desired. In addition, the inclusion of functional groups that enhance adsorption properties would improve filtration performance. Biochar does not have a very high surface area and it tends to have a large fraction of macro ($d > 50\,nm$) and meso ($2 < d < 50\,nm$) porous as opposed to microporosity, making it good for soil applications but not for filter material.

Biochar can be post-processed by chemical or physical activation in order to modify its porosity distribution and increase surface area and adsorption properties. Physical activation involves subjecting biochar to high temperature steam or carbon dioxide (around $800\,°C$), while chemical activation involves soaking biochar with KOH or NaOH and then heating to temperatures around $400-500\,°C$. Activated carbon has a higher fraction of fixed carbon and its surface area and porous size distribution vary with burn-off percentage. An excellent source of information for activated carbon can be found in Marsh and Reinoso (2006).

1.5 Pyrolytic Graphite

As biomass is carbonized, its electrical conductivity increases by several orders of magnitude. In a similar manner, graphite is made when a mixture of hydrocarbons are heated to graphitization temperatures ($2000\,°C$ or higher). As opposed to biochar and activated carbon, graphite is a crystalline material with a comparatively low porosity. These attributes enhance the electrical conductivity of graphite, which is about two to three orders of magnitude higher than that of biochar obtained at temperatures above $700\,°C$. However, compared to metals, the electrical conductivity of graphite is more than two orders of magnitude lower. For example, the electrical conductivity of POCO graphite is 7.8×10^4 (Ω m)$^{-1}$ compared to copper, which is 3.58×10^7 (Ω m)$^{-1}$ (Entegris, 2015). In this chapter, properties of (polycrystalline) POCO® Graphite AXF-5Q are used for models shown in Chapter 5, so when future reference is made to graphite without a qualifier, it means POCO graphite. The thermophysical properties of POCO and other commercial polycrystalline graphites are nearly isotropic.

Tables 1.1–1.4 provide a list of properties for a graphite as well as activated carbon.

Bibliography

A.K. Bain and P. Chand. *Ferroelectrics: Principles and Applications.* Wiley-VCH Verlag GmbH & Co., 2017.

A.O. Balogun, O.A. Lasode, and A.G. McDonald. Thermo-physical, chemical and structural modifications in torrefied biomass residues. *Waste and Biomass Valorization*, 9:131–138, 2018.

J. Barker-Jarvis, M.D. Janezic, B. Riddle, C.L. Holloway, N.G. Paulter, and J.E. Blendell. Dielectric and conductor-loss characterization and measurements on electronic packaging materials. Technical Report NIST Technical Note 1520, National Institute of Standards and Technology, 2001.

P. Chen, Y. Cheng, S. Deng, X. Lin, G. Huang, and R. Ruan. Utilization of almond residues. *International Journal of Agricultural and Biological Engineering*, 3(4):1–18, 2010.

W. De Jong. Chapter 2: Biomass composition, properties, and characterization. *Biomass as a Sustainable Energy Source for the Future: Fundamentals of Conversion Processes.* Ed. W. De Jong and J.R. Van Ommen, Wiley and American Institute of Chemical Engineers, Inc., 36–68, 2014.

G. Diaz, N. Sharma, E. Leal-Quiros, and A. Munoz-Hernandez. Enhanced hydrogen production using steam plasma processing of biomass: experimental apparatus and procedure. *International Journal of Hydrogen Energy*, 40:2091–2098, 2015.

M.A. Dietenberger, D.W. Green, D.E. Kretschmann, et al. Wood handbook-wood as an engineering material. Technical Report FPL-GTRâ113, U.S. Department of Agriculture, Forest Service, Forest Products Laboratory, Madison, WI, March 1999.

C. Dupont, R. Chiriac, G. Gauthier, and F. Toche. Heat capacity measurements of various biomass types and pyrolysis residues. *Fuel*, 115:644–651, 2014.

Entegris. Properties and characteristics of graphite. Technical report, Entegris, Inc., January 2015. http://poco.com/Portals/0/Literature/Semiconductor/IND-109441-0115.pdf.

S. Gaur and T.B. Reed. *An Atlas of Thermal Data For Biomass and Other Fuels.* National Renewable Energy Laboratory, 1995.

W.P. Goss and R.G. Miller. Thermal properties of wood and wood products. Technical report, Thermal Performance of the Exterior Envelopes of Buildings V, Clearwater, FL, USA, 1992.

W.L. James. Dielectric properties of wood and hardboard: Variation with temperature, frequency, moisture content, and grain orientation. Technical Report FPL-245, U.S. Department of Agriculture, Forest Service, Forest Products Laboratory, Madison, WI, 1975.

M. Kumar and R.C. Gupta. Electrical resistivity of acacia and eucalyptus wood chars. *Journal of Materials Science*, 28(2):440–444, 1993.

J.H. Kwon, S.B. Park, N. Ayrilmis, S.W. Oh, and N.H. Kim. Effect of carbonization temperature on electrical resistivity and physical properties of wood and wood-based composites. *Composites Part B: Engineering*, 46:102–107, 2013.

J. Lehmann. Bio-energy in the black. *Frontiers in Ecology and the Environment*, 5, 2007. https://doi.org/10.1890/060133

H. Marsh and F.R. Reinoso. *Activated Carbon*. Elsevier, 2006.

H. McLaughlin. Characterizing biochars: attributes, indicators, and at-home tests. In *The Biochar Revolution: Transforming Agriculture & Environment*. Ed. P. Taylor, pages 89–111. Global Publishing Group, 2010. ISBN 978-1-921630-41-5.

A. Mianowski, M. Owczarek, and A. Marecka. Surface area of activated carbon determined by the iodine adsorption number. *Energy Sources, Part A: Recovery, Utilization, and Environmental Effects*, 29(9):839–850, 2007.

A. Munoz-Hernandez. *Charge and Joule Heat Transport in Carbonaceous Materials and Activation of Biochar*. PhD dissertation, University of California - Merced, 2018.

L.S. Parfen'eva, T.S. Orlova, N.F. Kartenko, B.I. Smirnov, I.A. Smirnov, H. Misiorek, A. Jezowski, J. Muha, and M.C. Vera. Structure, electrical resistivity, and thermal conductivity of beech wood biocarbon produced at carbonization temperatures below 1000 °C. *Physics of the Solid State*, 53(11):2398–2407, 2011.

K.W. Ragland, D.J. Aerts, and A.J. Baker. Properties of wood for combustion analysis. *Bioresource Technology*, 37(2):161–168, 1991.

R.J. Ross. *Wood Handbook: Wood as an Engineering Material*. U.S. Dept. of Agriculture, Forest Service, Forest Products Laboratory, Madison, WI, centennial edition, 2010.

C. Skaar. *Wood-Water Relations*. Springer-Verlag, 1988.

R.E. Sonntag, C. Borgnakke, and W.G.J. Van Wylen. *Fundamentals of Thermodynamics*. Wiley, 6 edition, 1998.

H. Sugimoto and M. Norimoto. Dielectric relaxation due to interfacial polarization for heat-treated wood. *Carbon*, 42(1):211–218, 2004.

R.H. White and M.A. Dietenberger. Wood products: thermal degradation and fire. In *Encyclopedia of Materials: Science and Technology*. Ed. W. De Jong and J.R. Van Ommen, pages 9712–9716. Elsevier, 2001.

T.T. Win and Y.M.S. Nang. Physical and dielectric properties of palm shell biochar. *International Journal of Engineering Research & Technology (IJERT)*, 6(5):213–215, 2017.

X. Yang, H. Wang, P.J. Strong, et al. Thermal properties of biochars derived from waste biomass generated by agricultural and forestry sectors. *Energies*, 10(469):1–12, 2017.

S. Yorgun and D. Yildiz. Preparation and characterization of activated carbons from paulownia wood by chemical activation with H_3PO_4. *Journal of the Taiwan Institute of Chemical Engineers*, 53:122–131, 2015.

S.L. Zelinka, D.S. Stone, and D.R. Rammer. Equivalent circuit modeling of wood at 12% moisture content. *Wood and Fiber Science*, 39(4):556–565, 2007.

2

Conventional Processing Methods

2.1 Biomass Processing

The world is currently experiencing a time of rapid changes that require resilient energy generation as well as adaptation to uncertain conditions. For instance, in 2018, the demand for energy in the world was predicted to grow by more than 25% until 2040 (IEA, 2018). Due to slower global economic growth in 2019, the global energy demand increased by less than half the rate of the previous year (IEA, 2020a). It can be said that the world was not prepared for the events of 2020, where the COVID-19 pandemic caused more disruption to the energy sector than any other event in recent history (IEA, 2020b), with impacts that will be felt for years and uncertain predictions of future demand and supply. In addition, the recent sixth assessment report from the Intergovernmental Panel on Climate Change concludes that widespread and rapid changes in the atmosphere, ocean, cryosphere, and biosphere have occurred unequivocally due to human influence in warming the atmosphere, ocean, and land (IPCC, 2021).

This situation accentuates the need to reduce carbon emissions released to the atmosphere.

Biomass processing technologies can be classified in a variety of ways, but a useful arrangement is by means of end product:[1]

- direct combustion (for power)
- anaerobic digestion (for biogas)
- fermentation (for alcohols)
- oil extraction (for biodiesel)
- pyrolysis (for biochar, gas, and oils)
- gasification (for synthesis gas)

---

1 https://www.eubia.org/cms/wiki-biomass/biomass-processing-technologies/.

Voltage-Enhanced Processing of Biomass and Biochar, First Edition. Gerardo Diaz.
© 2022 John Wiley & Sons Ltd. Published 2022 by John Wiley & Sons Ltd.

In this chapter, we will concentrate on processes that can adapt well to using voltages and currents to enhance biomass conversion performance. In general, the thermochemical conversion composed by (i) direct combustion, (ii) pyrolysis, or (iii) gasification can be improved by using thermal or nonthermal plasma, or by utilizing the effects of Joule heating. However, direct combustion is not considered a carbon negative process so it will not be considered in this work.

2.1.1 Biomass Pyrolysis

Pyrolysis is the process of heating a carbon-bearing solid material under oxygen-starved conditions (Lehmann and Joseph, 2009). Although there are small fractions of organic extractives and inorganic minerals, biomass is mostly composed of cellulose, hemicellulose, and lignin, with these components thermally decomposing at different temperatures and rates during pyrolysis. The products of the decomposition include fractions of solid, liquid, and gases. The solid product is referred to as *biochar* and is a relatively porous material that can be used for soil amendment or for filtering of some pollutants. As biomass is heated, volatiles in the form of vapors are produced, which can be condensed into liquid form. In addition, noncondensable gases such as H_2, CO, CO_2, CH_4, and other molecules leave the reactor as product gases. Depending on parameters such as heating rate and residence time of the feedstock in the reactor, the process can be categorized as slow, fast, or flash pyrolysis. The ranges of operation for these three modes of pyrolysis vary in the literature, but the heating rate, temperature, and residence time are usually the main parameters reported in most analyses. It is observed that for low operating temperatures in the range between 200 and 300 °C, with heating rates below 50 °C/min and for very long residence times from hours to days, the process is usually referred to as *torrefaction*, which is a very slow pyrolysis process performed to generate a stable material that can be used as fuel. About 90% of the energy content is retained and the hydrophobic material produced can be stored for a long time.[2]

Table 2.1 shows typical ranges of parameters for the different types of pyrolysis. Slow pyrolysis operates at temperatures above torrefaction but below fast pyrolysis. Resident times tend to be moderate in the range between 5 and 45 minutes with low heating rates. The main purpose of slow pyrolysis is to produce biochar, so a low fraction of gases is obtained, which in some reactors is utilized to provide part of the heat required by

---

2 https://www.e-education.psu.edu/egee439/node/537.

Table 2.1 Pyrolysis methods

Method	Temperature (°C)	Residence time	Heating rate (°C/s)	Products
Slow	300–500	5–45 min	10	Biochar/bio-oils/gases
Fast	400–650	0.5–2 s	10–200	Bio-oils/gases/biochar
Flash	700–1000	< 0.5 s	> 500	Gases/bio-oils

Source: Based on Hu and Gholizadeh (2019).

the process. Temperatures in the 400–500 °C with low residence time favor the conversion from biomass into liquid products, such as bio-oils, mainly composed of oxygenated compounds (Hu and Gholizadeh, 2019), while gases and bio-oils are the main product of flash pyrolysis. The production of biochar involves a complex process that involves large number of reactions that depend on the feedstock characteristics as well as the operating conditions. However, reactions can be classified as:

- *Primary*: Where (i) *charring*, i.e. char formation occurs together with (ii) *depolymerization*, where cracking of bond linkages happens generating volatile compounds and gases, and (iii) *fragmentation*. *Primary biochar* forms as devolatilized biomass leaves behind a carbonaceous residue.
- *Secondary*: Where compounds formed during primary reactions undergo *cracking* or *recombination* reactions. Cracking forms lighter compounds, whereas recombination forms heavier compounds that can deposit on the char surface. *Secondary biochar* forms from the decomposition of organic vapors to form coke (Lehmann and Joseph, 2009).

In order to maximize biochar production, low heating rates are utilized, which results in a sequence of processes that involve drying, where most of the water is removed at temperatures between 100 and 120 °C. As temperature is increased between 200 and 300 °C, some of the biomass chemical bonds start to break down through endothermic reactions that produce methanol, acetic acid, and volatile organic compounds (VOCs). At around 320 °C, small molecules such as H_2, CO, CH_4, and CO_2 start to form, where large polymers start decomposing, and some of the oxygen contained in the biomass is released allowing for exothermic oxidation reactions to take place (Taylor and Mason, 2010). This is very important, especially for large pyrolysis reactors where nonuniform temperature distribution might generate hot spots that promote energy-releasing oxidation reactions increasing the temperature even more, making it very hard to control the overall biomass conversion process. If additional volatile compounds need to be removed, the temperature needs to be raised beyond 600 °C, which requires additional external heat.

2.1.2 Biomass Gasification

Gasification is a thermochemical process in which reactions between a carbonaceous material and the gasification agent take place and a product gas is generated (Zhang et al., 2019). In the case of biomass gasification with air, the product gas contains mainly H_2 and CO (a mixture referred to as synthesis gas or syngas), in addition to other components such as CO_2, CH_4, O_2, N_2, and small fractions of light hydrocarbons. Gasifier sizes range from a few kilograms of biomass to several tons per hour, and they are mainly used to generate power/heat or liquid fuels. Figure 2.1 provides a schematic of the main different types of gasifiers. In an *updraft gasifier*, the biomass is introduced from the top and the air from the bottom in counter flow configuration. Combustion occurs at the bottom with the area right above it being reduction zone, as shown in Figure 2.1a. Hot gases heat up the biomass in the pyrolysis zone and the extra heat available helps to dry the top zone where new biomass material is added. Because the pyrolysis zone generates VOCs, the product gas from these gasifiers has a high content of tars, making this type of gasifier not desirable for power generation from an internal combustion engine.

A *downdraft gasifier* is a common type of reactor that can be scaled to generate 1–2 MW of electric power. Biomass is fed from the top and air is injected

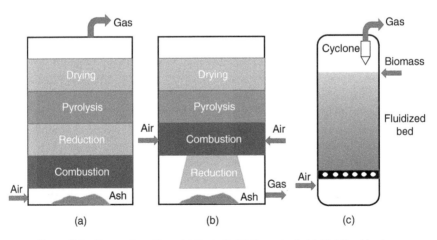

Figure 2.1 Types of gasifiers. (a) Updraft. (b) Downdraft. (c) Fluidized bed.

at the combustion zone as shown in Figure 2.1b. As shown in the figure, the biomass is first dried with some of the heat losses from the pyrolysis zone. Then the feedstock undergoes pyrolysis where VOCs are released and combusted at temperatures between 800 and 1100° C. The remaining char reacts with carbon dioxide and steam to produce CO and H_2 at the reduction section. Although the product gas is cleaner than in the updraft gasifier, power plants require a number of cleaning stages that involve cyclone separators, water and oil scrubbers, and filters before the gas can be injected to a generator set. For large units, biochar is collected at the bottom and removed on a regular basis. Depending on the type of operating conditions, this biochar could have a high ash content. However, due to the high operating temperature, the surface area of the biochar produced is higher than for typical pyrolysis reactors, making this biochar more appropriate for applications that require filtering of pollutants.

A third type of gasifier is the *fluidized bed* reactor where biomass particles are suspended in an oxygen-rich gas where good mixing occurs. In order for the particles to behave like a fluid, small sizes ($< 6\,\text{mm}$) need to be used, which allows for high carbon conversion rates (90–95%). The high temperature operation tends to decompose most of the tars, oils, phenols, and other liquid byproducts. There are several other types of gasifiers, but not all of them are suitable for biomass conversion. A very detailed description of different types of gasifiers and their performance can be found in NETL (2002, 2006).

2.2 Biochar Production and Post Processing

The different types of thermochemical conversion processes for biomass are classified as combustion, gasification, pyrolysis, and hydrothermal processing. While combustion and gasification are usually used for power and process heat generation, hydrothermal conversion is used to produce chemicals such as hydrogen and methane or bio-crude (Kruse and Dahmen, 2018). Fast pyrolysis is utilized to obtain bio-oils, although slow pyrolysis can also be used to produce natural pesticides, herbicides, and insecticides (Hagner et al., 2020).

As the market for biochar continues to develop, there is a growing interest in producing this material at a low cost. Therefore, gasification reactors for which biochar is a byproduct of large-scale power or process-heat generation constitute a feasible source of this material. Another source of biochar involves material produced in advanced units specifically designed for biochar production, where a screw reactor is utilized to move biomass

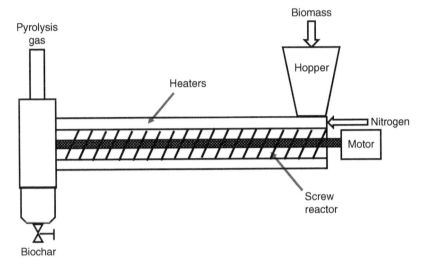

Figure 2.2 Screw reactor for biochar production.

continuously as heaters raise the temperature to pyrolysis conditions, as shown in Figure 2.2. Some of these units can utilize the heating value of the pyrolysis gas to inject it back into the furnace to lower the cost of external fuel used. Because these systems need to be sealed to prevent air from reaching the biomass, they are expensive and are not intended for processing biomass at small scale production.

On the other hand, there is a simple and low-cost reactor design that provides external heat for the pyrolysis process and at the same time keeps the air from interacting with the biomass. These reactors are called two-barrel nested biochar retort, also known as *retorts*. The designs are depicted in Figure 2.3 where an inverted 30-gallon (0.113 m³) drum is filled with biomass, and it is placed inside a 55-gallon (0.208 m³) drum which is partially filled with wood. Small openings for inlet primary air are made at the bottom of the large drum. The external wood is lit and as it burns, heat is provided to the biomass inside the smaller inverted drum. As volatiles are released from the internal biomass, they can only escape from the bottom of the inverted drum so they burn together with the external wood. The process ends when all the external wood has been consumed and there are no more volatiles being released from the internal biomass. Secondary air can be added at the top of the large drum to reduce emissions of particulate matter and unburned hydrocarbons.

It is evident that keeping the reactor sealed from air and burning an external fuel to generate heat for the pyrolysis process is expensive, so many

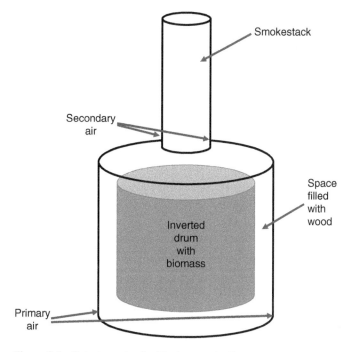

Figure 2.3 Retort reactor for biochar production.

biochar reactors use a small amount of air to generate partial oxidation reactions in the biomass to provide the necessary pyrolysis heat, at typical equivalence ratios between 20% and 25%, where the *fuel–air* ratio is given by the expression:

$$F/A = \frac{\text{mass of fuel}}{\text{mass of air}} \tag{2.1}$$

and the *equivalence ratio* is the ratio of the actual fuel–air ratio to the fuel–air ratio for complete combustion of biomass (stoichiometric):

$$\Phi = \frac{F/A}{(F/A)_{stoic}} \tag{2.2}$$

Within this category of biochar reactors that use air to generate heat through exothermic reactions, the *Pit kiln* has a stack of smoldering wood buried underground and covered by a layer of branches and leaves. Openings on the ground allow air to enter the pit to keep the process in operation. Depending on the size of the pit, the conversion can take up to several days to be finished, usually providing low biochar yields and nonuniform quality. Since only primary air is provided, a significant amount of particulate matter and VOCs are emitted to the atmosphere.

A similar design but above the ground is the *Mound kiln* where wood is stacked vertically with an orifice in the middle and covered by leaves and a layer of dirt. Several openings are located at the bottom of the kiln to allow air to enter the wood area. Performance is similar to the pit kiln with considerable air pollution being generated and low biochar yield (Lehmann and Joseph, 2009). A number of kiln designs have become commercially available that can process biomass of different sizes.

As indicated by the California Air Resources Board's website: "Air pollution harms people's health, damages agricultural crops, forests, ornamental, and native plants, and creates the haze that reduces visibility." Therefore, technologies that include secondary air combustion or air pollution control systems through catalytic converters or thermal oxidizers, provide an additional benefit to biochar production.

A simple and low cost design that utilizes secondary air is the *Top-Lit UpDraft* or TLUD reactor (Anderson, 2010). Figure 2.4 shows an operating unit, where low equivalence ratio air is injected at the bottom of a 55-gallon ($0.208 \, m^3$) drum filled with biomass. The top of the biomass material is lit usually by pouring a small amount of flammable liquid and using a lighter or propane torch. Once the top layer of the material is burning, an exhaust stack is placed on top of the drum, which has openings for the secondary air to combust pyrolysis gas produced at the reactor. It is observed that the effect of the secondary air significantly reduces the release of smoke to the atmosphere.

Figure 2.4 Top-Lit UpDraft or TLUD biochar generator.

There are a number of commercially available units that use air to generate exothermic reactions that pyrolyze the feedstock. Some require more diagnostic instrumentation than others. A more complex design that produces consistent results and still uses external air to provide the heat required for pyrolysis exposes the biochar to a processing temperature and oxygen level below that of luminous combustion, while the oxygen in the vapor space surrounding the biochar is controlled to promote oxidation reactions that generate heat (McLaughlin, 2016). The material is continuously rotated which increases the uniformity of the biochar produced. Gases generated are evacuated from the reactor and burned using secondary air. The process can be scaled to produce from $0.5 \, yd^3$ $(0.38 \, m^3)$ per batch to $20 \, yd^3$ $(15.29 \, m^3)$ per batch.

For removing very large pieces of wood, especially for forest waste, there are container-size conversion units that utilize an air curtain above the combustion zone, that prevents the volatiles and other pollutants to be released to the atmosphere. They can reduce debris volume by 90% and can be transported directly to the site where the feedstock is located. Due to the large size and nonuniform temperature distribution within the large processing volume, the biochar produced is of nonuniform quality.

2.2.1 Biochar Activation

Biochar is mainly considered as a carbonaceous material that is obtained from thermochemical conversion of biomass with the purpose of using it as soil amendment. However as production of this material grows, it is unclear if the market for soil amendment will become saturated with the larger production of biochar. In addition, it has become apparent that commercial success in gasification, particularly in the high-cost forested areas of California, depends on the sale of the biochar byproduct as much as the sale of electricity. With more biomass gasification plants being constructed, a rapid expansion of biochar supply is forcing the industry to create new markets for this byproduct. The production of activated carbon from biochar constitutes a possible route where additional industrial markets can be developed, targeting materials suitable for liquid or air-pollution filters. The next Sections (2.2.1.1–2.2.1.5) provide information about production and characterization of activated carbon produced from local biochar.

2.2.1.1 Production of Activated Carbon

Activated carbon is a highly porous material with surface functional groups that enhance the adsorption of a given adsorbate (pollutant) from a given aqueous solution or gas. Commercially activated carbon is mainly produced from two sources: coal or biomass. The former is produced in the United States from various types of coal, while the latter is produced from coconut shells – mainly in Asian countries.

At laboratory scales, biochar is typically activated using two methods: chemical activation or physical activation (Marsh and Reinoso, 2006; Ioannidou and Zabaniotou, 2007). In this section, we will concentrate on physical activation with steam.

There are four key parameters involved in steam activation of a given biochar: (i) activation temperature, (ii) activation time, (iii) steam/water flow rate, and (iv) particle size. The particle size usually covers a range from 0.6 to 2.36 mm because particles that are too small can be carried over by the steam flow, and because much of the inner space will not become activated when particles that are too big.

It has been found that at temperatures below 700 °C the porosity and surface areas are relatively low (Azargohar and Dalai, 2005, 2008; Demiral et al., 2011). On the other hand, at temperatures above 850 °C, more than 50% of the biochar burns off, so that the higher surface area comes at the expense of higher mass loss. Moreover, it has been shown that for temperatures around 800 °C, large surface areas and pore volumes are generally obtained for various types of feedstock (Aworn et al., 2008; Demiral et al., 2011; Alvarez et al., 2015). Activation duration times range from about 15 to 60 minutes (Demiral et al., 2011; Alvarez et al., 2015), where the steam flow rates reported in recent articles range from about 1 to 8 g/min (Aworn et al., 2008; Azargohar and Dalai, 2005; Demiral et al., 2011). For the results shown here, steam flow rate was kept in this range, while the activation temperature remained near 800 °C, and the residence times varied between 15 and 75 minutes, as long as the total mass loss was kept within 50% (Munoz-Hernandez et al., 2018).

2.2.1.2 Experimental Setup

A schematic of the experimental setup for superheated steam activation is shown in Figure 2.5. A vertical tube furnace with a maximum operating temperature of 1200 °C and 24 in. of heated length (60.96 cm) is used. A high temperature stainless steel (310S) work-tube with an inner diameter (ID) of 5.10 cm and flanges at both ends is placed inside the furnace. Inside the work-tube, a 2 in. (5.08 cm) outer diameter (OD) coil made with 3 mm OD 310S stainless steel tube extends from the beginning (bottom) of the heated region of the furnace to approximately 45.7 cm up into the heated region. It then connects to a 2 in. (5.08 cm) OD reactor, where the biochar

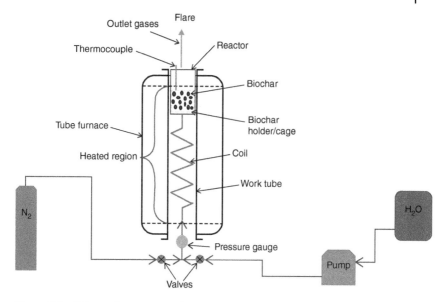

Figure 2.5 Schematic representation of the biochar activation experimental setup.

is placed during the activation process. The reactor is 18 in.-long (45.7 cm) and sits at the top of the work tube. It extends down to about 7 in. (17.78 cm) into the heated portion of the furnace, where it connects to the coil. The reactor has an end-cap at the top, into which an 18 in. (45.7 cm) long thermocouple is inserted to measure the actual temperature near the biochar. In addition, a foot long (30.48 cm), 1/4 in. (0.635 cm) pipe is connected to the reactor end-cap. Gases coming out from the activation process are flared at the end of the pipe. At the inlet (bottom) of the coil, a 1/4 in. (0.635 cm) OD stainless steel tube is attached, which extends well below the bottom of the work tube. Toward the bottom end of the 1/4 in. (0.635 cm) OD tube, a pressure gauge is connected and then the tube ends at a tee. One end of the tee receives a nitrogen line, while the other end receives a water line. The water line is fed by a high-pressure low-flow water pump. Deionized water is used for the experiments, which is stored in a container to feed the pump. On the other side of the tee, the nitrogen is fed from a nitrogen tank.

2.2.1.3 Procedure

Biochar Sample Preparation Peach pit biochar was selected due to its local availability. This is a byproduct produced from biomass gasification at a

0.5 MW power plant. It was sieved to a size range between 0.6 and 2.36 mm, and dried in an oven for eight hours at 105 °C. A sample of 20 g of the sieved and dried biochar was used for each activation run.

Physical Activation Procedure   Industrial scale physical activation of biochar occurs in large kilns where the input biochar is transported in conveyor belts and dropped inside the reactor which operates between 800 and 1000 °C. Therefore, the material experiences a very high rate of heating from near ambient conditions to the temperature of the activation process. Our laboratory procedure intends to mimic this large heating rate by placing a biochar sample at ambient temperature inside a furnace operating at steam activation temperature. The experimental procedure is as follows: the furnace is preset and heats up to the activation temperature at a maximum rate of 8 °C/min. Once the furnace reaches the activation temperature, the water pump is turned on at the desired flow rate to initiate the production of superheated steam. After about 30 minutes, the reactor where the biochar sample is placed, also reaches the set furnace temperature. This happens since the reactor temperature always lags the electrically heated furnace temperature due to thermal inertia. At this time, the water pump is turned off, the reactor end-cap is taken off, and the biochar sample is quickly placed inside the reactor. Then the reactor is closed and the pump is switched back on. Also, a propane torch is fired up to flare the outflowing gases. All these steps are performed in approximately 30 seconds. Because the thermocouple is attached to the end-cap, which is removed temporarily while inserting the biochar into the reactor, the temperature reading falls well below the furnace temperature, however, as soon as the thermocouple is reinserted, the temperature reading goes back to the furnace temperature and eventually reaches the same value. The temperature profile is recorded for all the experiments, and a log-mean-temperature is used to refer to an average activation temperature, which is close to the set furnace temperature. Upon completion of the activation process, the water flow is stopped. At this time, nitrogen flow is initiated, and runs continuously thereafter while the activated carbon sample cools down to a temperature below 200 °C. The sample is taken out of the reactor at low temperatures to avoid instantaneous combustion of the biochar.

This process replicates what is done in industry more closely. Alternatively, the biochar could be placed in the reactor when the furnace is turned on, and nitrogen could flow during the heating process, as is typically done in other published laboratory experiments.

2.2.1.4 Performance Evaluation

The ultimate and proximate analyses for raw biochar and activated carbon sample are shown in Table 2.2. The sample was activated with a furnace temperature of 800 °C and a steam flow rate of 0.8 g/min for 30 minutes.

The material loss due to the steam activation process is usually referred to as burn off percentage given by Eq. (2.3),

$$\text{Burn off} = \frac{\text{initial mass} - \text{final mass}}{\text{initial mass}} \times 100 \ (\%) \tag{2.3}$$

Figure 2.6 shows the burn off percentage with respect to activation time for peach pit biochar samples exposed to different operating temperatures and steam flow rates. Operating temperatures range between 800 and 850°C and the mass flow rates of steam are in the range between 0.8 and 4.3 g/min. It is observed that high burn off percentages were obtained at high mass flow rates of steam, high temperature, and longer activation times. High loss of mass is economically detrimental for large-scale production of activated carbon. Figure 2.7 shows the BET surface area analysis for the treated biochar samples and one raw biochar sample. It is observed that the raw biochar sample (*) has a surface area of less than 1 m²/g, which shows the low capability of being used as a filtering material. On the other hand, BET surface areas near 600 m²/g were obtained with 810 °C and a mass flow rate of steam

Table 2.2 Properties of raw and activated peach pit biochar

	Raw biochar	Activated biochar
Proximate analysis (wt.%, dry basis)		
Ash	21.24	35.00
Volatile matter	23.78	5.04
Fixed carbon	54.98	59.96
Ultimate analysis (wt.%, dry basis)		
Carbon	65.62	63.17
Hydrogen	2.33	<0.10
Nitrogen	0.57	0.32
Oxygen (by difference)	10.12	1.44
Chlorine	0.10	0.05
Sulfur	0.02	0.02

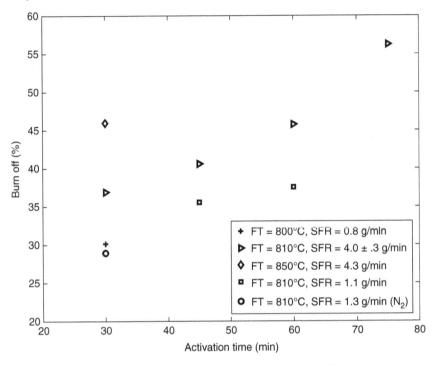

Figure 2.6 Burn off % vs. activation time for different steam flow rates and furnace temperatures. Legend: FT: Furnace temperature; SFR: Steam flow rate; N$_2$: Nitrogen.

of 4.3 g/min for activation times of 40 and 70 minutes, shown in ▷. It is noted that high flow rate of steam and higher operating temperatures translate into higher energy consumption to produce activated carbon. The figure also shows a circle marker that corresponds to the BET surface area for a case where low heating rate was utilized on the biochar sample by slowly reaching steam activation temperatures with the biochar sample inside the reactor under nitrogen flow. Once the steam activation temperature was reached, the nitrogen flow was stopped and steam flow was started. The aspect of energy consumption to produce activated carbon is shown in Figure 2.8, where the BET surface area has been multiplied by the fraction of mass left after the activation process and divided by the energy used to produce the treated biochar sample. It is observed that the samples with high steam flow rate and activation times are at the bottom of the group, which means that it is very energy intensive to produce these samples. On the other hand, even though the + marker, produced with 800 °C, 0.8 g/min of steam mass flow

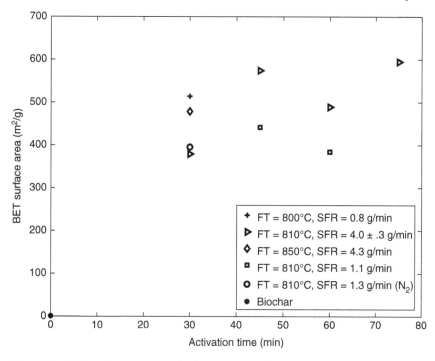

Figure 2.7 BET surface area vs. activation time for different steam flow rates and furnace temperatures. Legend: FT: Furnace temperature; SFR: Steam flow rate; N_2: Nitrogen.

rate, and 30 minutes of activation time, reached only 513 m²/g of BET surface area, it is a good compromise between surface area per energy used in the process.

The SEM picture of the sample with a + marker is shown in Figure 2.9 which characterizes the surface geometry that features a mean pore size of 20.7 Å and a total pore volume of 0.27 cm³/g. It is noted that the burn off percentage of this sample was only 30.2%.

2.2.1.5 Chemical Activation

Chemical activation consists of soaking biochar with strong chemicals such as sodium hydroxide, potassium hydroxide, zinc chloride, phosphoric acid, or potassium chloride. The material is then heated to temperatures below 500 °C where the remaining sodium, potassium, etc. is then washed away. Recent tests with biochar from a gasification plant for power generation soaked in NaOH or KOH and then dried, showed lower performance in removing hydrogen sulfide, benzene, VOCs, and siloxane compared to

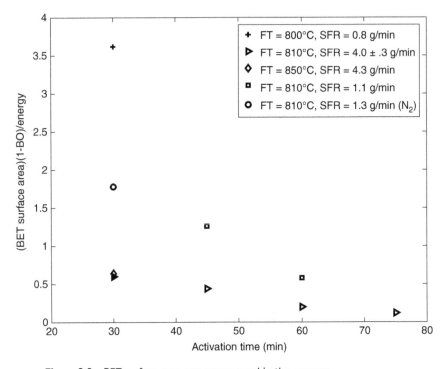

Figure 2.8 BET surface area per energy used in the process.

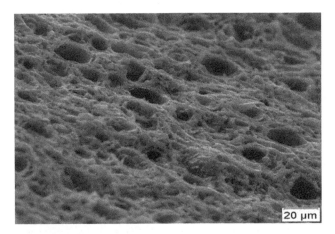

Figure 2.9 Scanning electron microscope image of activated biochar sample.

commercial granular activated carbon (GAC) when tested at industrial waste water treatment plants (Kester, 2021). Laboratory tests for H_2S breakthrough of this material using ASTM D6646 standard also showed lower performance compared to commercial GAC. More research needs to be conducted to improve performance of the locally produced material.

More detailed information for physical and chemical activation of biochar can be found in Ioannidou and Zabaniotou (2007).

Bibliography

J. Alvarez, G. Lopez, M. Amutio, J. Bilbao, and M. Olazar. Physical activation of rice husk pyrolysis char for the production of high surface area activated carbons. *Industrial and Engineering Chemistry Research*, 54(29):7241–7250, 2015.

P.S. Anderson. Making biochar in small gasifier cookstoves and heaters. In *The Biochar Revolution: Transforming Agriculture & Environment*. Ed. P. Taylor, pages 153–182. Global Publishing Group, 2010. ISBN 978-1-921630-41-5.

A. Aworn, P. Thiravetyan, and W. Nakbanpote. Preparation and characteristics of agricultural waste activated carbon by physical activation having micro-and mesopores. *Journal of Analytical and Applied Pyrolysis*, 82(2):279–285, 2008.

R. Azargohar and A.K. Dalai. Production of activated carbon from Luscar char: experimental and modeling studies. *Microporous and Mesoporous Materials*, 85(3):219–225, 2005.

R. Azargohar and A.K. Dalai. Steam and KOH activation of biochar: experimental and modeling studies. *Microporous and Mesoporous Materials*, 110(2):413–421, 2008.

H. Demiral, I. Demiral, B. Karabacakoğlu, and F. Tümsek. Production of activated carbon from olive bagasse by physical activation. *Chemical Engineering Research and Design*, 89(2):206–213, 2011.

M. Hagner, K. Tiilikkala, I. Lindqvist, K. Niemela, H. Wikberg, A. Kalli, and K. Rasa. Performance of liquids from slow pyrolysis and hydrothermal carbonization in plant protection. *Waste and Biomass Valorization*, 11:1005–1016, 2020.

X. Hu and M. Gholizadeh. Biomass pyrolysis: a review of the process development and challenges from initial researches up to the commercialisation stage. *Journal of Energy Chemistry*, 39:109–143, 2019. ISSN 2095-4956. https://doi.org/10.1016/j.jechem.2019.01.024.

IEA. World energy outlook 2018. Technical report, International Energy Agency, Paris, 2018. https://www.iea.org/reports/world-energy-outlook-2018.

IEA. Global energy review 2019. Technical report, International Energy Agency, Paris, 2020a. https://www.iea.org/reports/global-energy-review-2019.

IEA. World energy outlook 2020. Technical report, International Energy Agency, Paris, 2020b. https://www.iea.org/reports/world-energy-outlook-2020.

O. Ioannidou and A. Zabaniotou. Agricultural residues as precursors for activated carbon production - a review. *Renewable and Sustainable Energy Reviews*, 11(9):1966–2005, 2007.

IPCC. *Climate change 2021: The physical science basis.* Contribution of working group I to the sixth assessment report of the intergovernmental panel on climate change. Technical report, Intergovernmental Panel on Climate Change, Cambridge University Press, 2021.

G. Kester. Laboratory testing of forest biochar treatments for the wood based biochar as an alternative adsorption material for the control of off-gasses at wastewater treatment plants, 2021.

A. Kruse and N. Dahmen. Hydrothermal biomass conversion: Quo vadis? *The Journal of Supercritical Fluids*, 134:114–123, 2018. ISSN 0896-8446. https://doi.org/10.1016/j.supflu.2017.12.035. 30th Year Anniversary Issue of the Journal of Supercritical Fluids.

J. Lehmann and S. Joseph. *Biochar for Environmental Management: Science and Technology*. EarthScan, London, 2009.

H. Marsh and F.R. Reinoso. *Activated Carbon*. Elsevier, 2006.

H. McLaughlin. Method of increasing adsorption in biochar by controlled oxidation, 2016.

A. Munoz-Hernandez, S. Dehghan, and G. Diaz. Physical (steam) activation of post-gasification biochar derived from peach pits. In *ASME 2018 International Mechanical Engineering Congress and Exposition*. American Society of Mechanical Engineers, 2018.

NETL. Major environmental aspects of gasification-based power generation technologies - final report. Technical report, SAIC Prepared for NETL Gasification Technologies Program, 2002.

NETL. Different types of gasifiers and their integration with gas turbines, 2006.

P. Taylor and J. Mason. Biochar production fundamentals. In *The Biochar Revolution: Transforming Agriculture & Environment*. Ed. P. Taylor, pages 113–131. Global Publishing Group, 2010. ISBN 978-1-921630-41-5.

Y. Zhang, Y. Cui, P. Chen, S. Liu, N. Zhou, K. Ding, L. Fan, P. Peng, M. Min, Y. Cheng, Y. Wang, Y. Wan, Y. Liu, B. Li, and R. Ruan. Chapter 14 - Gasification technologies and their energy potentials. In *Sustainable Resource Recovery and Zero Waste Approaches*. Ed. M.J. Taherzadeh, K. Bolton, J. Wong, and A. Pandey, pages 193–206. Elsevier, 2019. ISBN 978-0-444-64200-4. https://doi.org/10.1016/B978-0-444-64200-4.00014-1.

3

Introduction to Plasmas

3.1 Thermal Plasmas

Plasmas are composed of a mixture of electrons, ions, and neutral species in local electrical neutrality. Electrons accelerated by an applied electric field, collide with larger particles, transferring kinetic energy, which contributes to the excitation and ionization of atoms and molecules, and to the heating of the gas (Samal, 2017). As a result, a highly conductive gas is formed, which allows for the flow of current through it. Thermal plasmas are typically produced with direct current (DC) transferred arcs, plasma torches, or radio frequency (RF) inductively coupled discharges. One of the main characteristics of thermal plasmas is that the kinetic temperature of the neutral particles, ions, and electrons are in local thermodynamic equilibrium (LTE), so they can be characterized by a single temperature. However, some deviations from LTE may exist, especially in the presence of steep temperature gradients (Boulos, 1991). Industrial applications of thermal plasmas tend to operate at pressures $p \geq 1$ atm, with electron densities in the range between 10^{22} and 10^{25} electrons/m^3, and electron kinetic temperatures between 0.25 and 1 eV (3000 to 11,600 K). In general, industrial plasmas have only a fraction of the total number of particles in the gas ionized.

Two typical configurations involving thermal arcs are shown in Figure 3.1. In a transferred arc, shown in Figure 3.1a, the cathode is in contact with the orifice gas (or working gas) which is ionized and forms the arc that exits the nozzle. The arc extends to the work section which acts as the positive electrode and thus the arc is transferred from the cathode to the work section. In applications such as welding, a shield gas is used to avoid oxidation of the metal being welded; however, in biomass conversion, the shield gas is not necessarily required. As shown in Chapter 1, biochar has a higher conductivity than biomass and thus, it can act as an anode without the need

Voltage-Enhanced Processing of Biomass and Biochar, First Edition. Gerardo Diaz.
© 2022 John Wiley & Sons Ltd. Published 2022 by John Wiley & Sons Ltd.

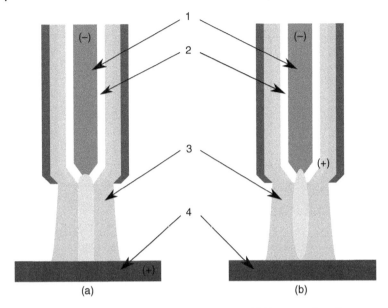

Figure 3.1 Thermal plasmas. (a) Transferred arc. (b) Non-transferred arc.
(1) Electrode (cathode). (2) Orifice gas. (3) Shield gas. (4) Work.

to significantly increase the applied voltage. Biomass can also act as an electrode, but much higher voltages are needed.

Figure 3.1b shows a non-transferred arc plasma torch where the enclosure of the orifice gas acts as the anode. An arc is formed between the negative and positive electrodes, which is pushed out of the nozzle by the gas flow rate. The work section does not act as an electrode in this case and shield gas is usually not required for biomass applications. An advantage of this type of devices is that they can heat up electrically nonconductive material, such as biomass (Tanaka, 2017). The flow of the gases helps to remove part of the heat from the walls of these two devices, but due to the intense heat generated by the arc, many torch designs circulate a fluid such as water to cool down wall temperatures. Figure 3.2 shows the exhaust of a plasma torch using air plasma in contact with wood sawdust (Foret, 2009).

Different methods can be used to start the electric arc. A simple mechanism is "drawing the arc" (Roth, 1995), where the cathode and anode are placed in contact with each other and a current is allowed to flow through them. Once the current is established, the electrodes are slowly separated from each other forming the arc. The current is reduced and the voltage increases, but the power remains almost the same. Extending the arc to a length that surpasses the capacity of the power supply will

Figure 3.2 Processing of biomass with a Patented Plasma ArcWhirl© Torch provided by Foret Plasma Labs, LLC.

result in extinguishing the arc. A different approach is to increase the voltage between both electrodes until electrical breakdown occurs and glow to arc transition occurs. Depending on the size of the gap between the two electrodes, this method might require high voltages to initiate the discharge, which will place a significant load on the power supply once high current starts to flow. Finally, external devices such as induction coils or spark generators can also be used to initiate the arc.

3.1.1 Mathematical Model

As mentioned in Section 3.1, the neutral particles, ions, and electrons are in LTE in thermal arcs, and this characteristic is used to model their behavior using a single temperature. For the purpose of modeling, the positive column of thermal arcs, i.e. the section of the arc excluding the zones near the electrodes, is usually considered as an axisymmetric cylinder, where all the heat in the plasma energy balance is due to Joule heating and the loses occur in the radial direction, where the arc is stabilized by the wall temperature. The temperature distribution can be obtained using the *Elenbaas–Heller* equation.

$$\frac{1}{r}\left(rk(T)\frac{dT}{dr}\right) + \sigma(T)E^2 = 0 \tag{3.1}$$

with boundary conditions $\frac{dT}{dr} = 0$ at $r = 0$ and $T = T_w$ at $r = R$, where the thermal conductivity, $k(T)$, and the electrical conductivity, $\sigma(T)$, are functions of temperature, and the electric field, E, is constant. The presence of the two material functions, i.e. $k(T)$ and $\sigma(T)$, is usually simplified by introducing the concept of *heat conduction potential* (Mostaghimi et al., 2017) also known as the *heat flux potential* (Liao et al., 2016; Fridman and Kennedy, 2004),

$$S(T) = \int_{T_w}^{T} k(\mathcal{T})d\mathcal{T}; \quad k(T)\frac{dT}{dr} = \frac{dS}{dr} \tag{3.2}$$

where introducing dimensionless variables, $x = r/R$ and $\theta = S/S_c$, Eq. (3.1) is transformed to:

$$\frac{1}{x}\left(x\frac{d\theta}{dr}\right) + \zeta\sigma(\theta) = 0 \tag{3.3}$$

with boundary conditions $\frac{d\theta}{dx} = 0$ at $x = 0$ and $\theta = 0$ at $x = 1$, where $\zeta = \frac{E^2 R^2}{S_c}$, and S_c is the heat-flux potential at the center of the cylinder. The electric conductivity can be expressed as an Arrhenius-type function (Shaw, 2006): $\sigma = A \exp\left(\frac{-T_i}{T}\right)$, which can be transformed to:

$$\sigma(\theta) = B\exp(-\theta_i/\theta) \tag{3.4}$$

where $\theta_i = S_i/S_c$.

The electric field and current are related by Eq. (3.5):

$$E = \frac{I}{2\pi \int_0^R \sigma r\, dr} \tag{3.5}$$

which, by means of using the dimensionless quantities, it provides the expression to obtain θ_i (Liao et al., 2016):

$$\frac{3}{4}\theta_i \exp\left(\frac{3}{4}\theta_i\right) = \frac{3\pi}{\sqrt{2}}\frac{\sqrt{BS_i}}{I/R} \tag{3.6}$$

This is a transcendental equation that can be solved for θ_i as a function of the gas parameters, B and S_i, and the ratio of current and the arc radius. Using the parameters for the electric conductivity of helium in Table 3.1, and for an arc current of $I = 100$ A and a radius at the wall of $R = 0.005$ m, the solution of Eq. (3.6) gives $\theta_i = 3.129$, which together with the definition $\theta_i = S_i/S_c$ gives $S_c = 30,363.7$ W/m. Following the derivation in Liao et al. (2016), it can be shown that

$$EI \approx \frac{8\pi S_i}{\theta_i^2} \tag{3.7}$$

Table 3.1 Parameters for electric conductivity at 1 atm.

Gas	B (A/V m)	S_i (W/m)
Helium	57,000	95,000
Argon	16,000	3,300

Source: Based on Shaw (2006).

which gives an electric field of 2439.1 V/m. Substituting E, R, S_c, and Eq. (3.4) into Eq. (3.3), the governing equation can be solved for the distribution of θ along the radius of the plasma arc. Using the definition $S = S_c \theta$, the heat flux potential can be obtained, as is shown in Figure 3.3. In many cases, it is important to know the actual temperature distribution in the arc. Experimental data for the thermal conductivity of helium at high temperatures can be found in Dunn and Eagar (1990) and is shown in Figure 3.4. Knowing that $S(1) = 0$ and $T(1) = T_w$, and making use of the data for the thermal conductivity as a function of temperature, Eq. (3.2) can be used to calculate the values of the temperature distribution inside the plasma arc. Figure 3.5 shows the temperature distribution inside the plasma arc along

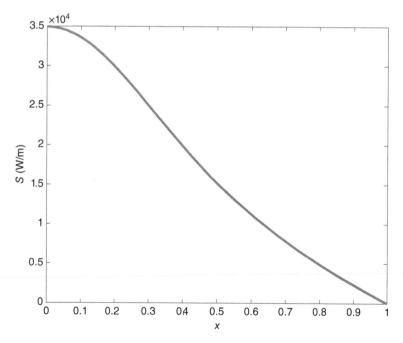

Figure 3.3 Heat flux potential for a helium arc with $I = 100$ A, and $R = 0.005$ m.

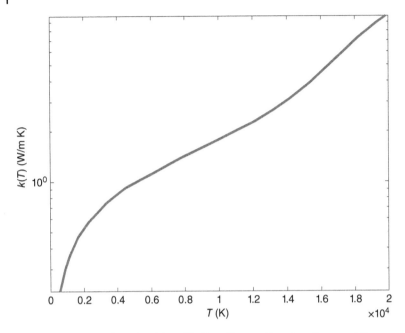

Figure 3.4 Thermal conductivity of helium. Source: Based on Dunn and Eagar (1990).

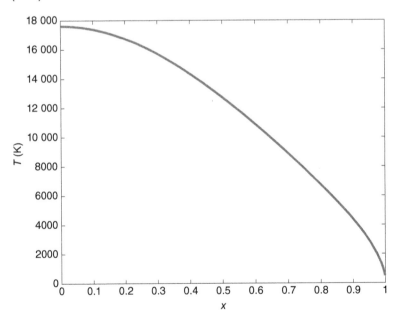

Figure 3.5 Temperature distribution inside the plasma arc.

the radius of the cylinder. In general, the Elenbaas–Heller approach is used for combinations of pressure and current where the plasma temperature does not exceed 1 eV (11,604 K) (Fridman and Kennedy, 2004); however, the plasma energy balance can be expanded to include radiation losses for such cases (Deron et al., 2006; Mostaghimi et al., 2017). For applications in biomass processing, plasmas are generally produced from air, steam, or carbon dioxide.

3.2 Nonthermal Plasmas

Thermal arcs are energy-intensive plasmas that are used in a wide range of applications, such as illumination, welding, and in the case of plasma torches, they have been used to destroy hazardous waste, vitrify inorganic materials, and to process biomass and municipal solid waste. In general, these processes rely on the intense heat emanated from these discharges. However, there are many applications where low-power plasma discharges are needed, for instance to modify surface properties, or to induce physical and chemical reactions within gases at relatively low temperatures (Liu et al., 1999). For this type of applications, nonthermal plasmas, also known as cold or nonequilibrium plasmas provide a wide range of operating conditions that generate a rich mix of reactive species, including excited molecules and atoms, radicals, and ions. Low-pressure plasma, also referred to as vacuum plasma has been applied in the electronics and semiconductor industries for printed circuit boards; however, this technology has also been introduced in the automotive, medical devices, and textile sectors (Lippens, 2007). Although the properties of low-pressure nonequilibrium plasmas have allowed these industries to continue to improve over time, vacuum plasmas are not as beneficial in the energy conversion sector due to large amount of material needed to be processed, the extra cost of the vacuum system (including the additional structural integrity of reactors), and the marginal profits per unit of mass processed. It is due to these reasons that atmospheric-pressure nonthermal plasmas are more commonly utilized in this and similar industries. The main parameters used to characterize these discharges are the electron, ion, and neutral gas number densities, as well as the electron, ion, and neutral gas temperatures. The electric field and corresponding current density are also necessary. The term *nonequilibrium plasma* comes from the fact that electron temperature is often higher than the ion and gas temperatures

$$T_e > T_i \approx T_n \tag{3.8}$$

Also, in many cases when the gas temperature is lower than typical combustion temperatures,

$$T_g < T_c \approx 1000\,^\circ C \tag{3.9}$$

the state is called *nonthermal plasma* (Yamamoto and Okubo, 2007).

Sections 3.2.1.1–3.2.5 describe some of the most common types of atmospheric pressure nonequilibrium electrical discharges used in industry.

3.2.1 DC Nonthermal Electrical Discharges

Atmospheric-pressure DC nonthermal plasma discharges can be classified into three main categories based on the level of current utilized.

3.2.1.1 Dark Discharge

Dark discharges utilize very low currents, i.e. of the order of micro Amperes or less, with relatively high voltages (several kilovolt). Due to the low number of excited species generated, and with the exception of certain ranges of operation of the more energetic corona discharge, the little excitation light produced is not visible (Roth, 1995), which explains the use of the term dark discharge. As the electric field is increased, electrons can gain sufficient energy to ionize neutral atoms, which generate more electrons that, in turn, ionize other neutral atoms. This process is known as electron avalanche and represents an exponential growth of the current, also referred to as *Townsend discharge*. The ionization process can be represented by the Townsend's first ionization coefficient, α, which is usually expressed in the form of the following exponential expressions, depending on the range of values of E/p:

$$\frac{\alpha}{p} = A \exp\left(-B\frac{p}{E}\right) \tag{3.10}$$

or

$$\frac{\alpha}{p} = C \exp\left(-D\left(\frac{p}{E}\right)^{1/2}\right) \tag{3.11}$$

where p is the pressure in Torr and E is in V/cm, and where parameters A, B, C, and D have been tabulated for several gases in Table 3.2 (Yuan et al., 2018).

As the voltage difference is increased nearing the electrical breakdown value, the dark discharge operates in *corona discharge* regime. Corona is usually generated near sharp surfaces, thin wires, sharp points, and surface asperities, where the electric field is high at one electrode but current is still low (mA or less). It can be generated from several configurations, which

Table 3.2 Parameters for first Townsend ionization coefficient.

Gas	A (cm Torr)$^{-1}$	B V(cm Torr)$^{-1}$	E/p V(cm Torr)$^{-1}$
He	3	34	20–100
Ne	4	100	100–400
Ar	12	180	100–600
H$_2$	5	130	150–600
N$_2$	8.8	375	27–200
Air	15	365	100–800
CO$_2$	20	466	500–1000
H$_2$O	13	290	150–1000
Gas	C (cm Torr)$^{-1}$	D V$^{1/2}$(cm Torr)$^{-1/2}$	E/p V(cm Torr)$^{-1}$
He	4.4	14	< 100
Ne	8.2	17	< 250
Ar	29.2	26.6	< 700

Source: Yuan et al. (2018).

include a sharp point to a spherical wall, sharp point to a flat wall, thin wire to a larger cylinder, or thin wire inside a flat-walled duct, among others. The discharge is bounded by the high-voltage electrode and the boundary of the active volume where the ionization process balances the electron attachment process (Chen and Davidson, 2002) in atmospheric air, as seen in Figure 3.6. Another known regime of corona is the *streamer mode*, where streamers are initiated from the negative glow sphere (Antao et al., 2009). Corona discharges have been industrially used in a number of applications, e.g. electrostatic precipitators to remove particles and pollutants from aerosols (Wen and Su, 2020), ionic wind generators (Monrolin et al., 2018; Colas et al., 2010), jet printers, electrohydrodynamics (EHD) thrusters (Gilmore and Barrett, 2015), ozone production, and for surface treatment (Chan, 1999). They can be positive or negative depending on the polarity of the high-voltage electrode, and display electron densities approximately between 10^8 and 10^{14} m^{-3} with a strong spatial nonuniformity, specially near the high-voltage electrode, and with electron temperatures in the range between 1 and 10 eV (11,000–110,000 K), while the ions and neutral gas remain near room temperature. Higher ionization rates and densities of

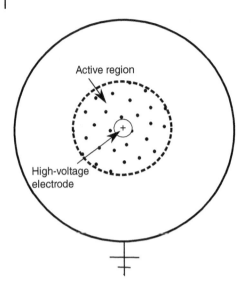

Figure 3.6 Schematic of a positive corona discharge.

excited species can result in visible light near the high-voltage electrode of these discharges.

3.2.1.2 Glow Discharge

The second category of DC discharges is referred to as glow discharges. There is a large volume of knowledge of glow discharges at low pressure; however, atmospheric pressure glow discharges (APGD) have become a topic of study in recent decades. In order to avoid the instabilities of electrons or glow-to-arc transition, a ballast resistor is placed in between the two electrodes (Li et al., 2008; Duan et al., 2005). An increase in current, for instance, by means of decreasing the ballast resistance connected in series with the discharge, can trigger the transition to *normal glow discharge*, where voltage is lower than for corona discharges and it remains mainly constant for a range of current values. A glow discharge is a self-sustained nonequilibrium discharge with a cold cathode that emits electrons mainly by secondary emission due to the effect of positive ions (Fridman and Kennedy, 2004). However, DC glow discharge at atmospheric pressure is hard to attain due to instabilities which lead to glow to arc transition (Staack et al., 2005). Some stabilization methods involve (i) pulsed discharge (Shao et al., 2018), where the voltage pulse duration is shorter than the time it takes a streamer to reach from one electrode to the other; or (ii) stabilization by fast gas flow along the anode surface (Hattori et al., 2006). Another important aspect of APGD corresponds to increasing its volume. Shao et al. (2017) used a rod array electrode to increase the volume of a spatially

Table 3.3 Parameters measured for APGD in air for 0.4 and 10 mA discharge current.

	0.4 mA	10 mA
Electrode spacing (mm)	0.05	0.5
Discharge voltage (V)	340	380
Discharge power (W)	0.136	3.8
Electric field (kV/cm)	5.0	1.4
Electron temperature (eV)	1.4	1.2
Electron density in negative glow (cm^{-3})	3×10^{13}	7.2×10^{12}
Electron density in positive column (cm^{-3})	—	1.3×10^{14}

Source: Modified from Staack et al. (2005).

diffused APGD. Typical characteristics of normal glow discharges are observed in air from low pressures to atmospheric values, which include a negative glow, Faraday dark space, and positive column region. Also, the current density remains constant even though the total current can be changing. Measured values of parameters for APGD in air are shown in Table 3.3.

As the current is increased to the point that the discharge covers the entire cathode, an increase in current involves an increase in current density, and an increase in voltage involves an increase in current which is referred to as *abnormal glow discharge*. Additional increase in current leads to the transition to nonthermal arc operation where the electron temperature is still different from the ion and neutral temperatures. At low pressures, the transition from glow-to-arc occurs at currents near 1 A.

3.2.2 Dielectric Barrier Discharge

For many applications, it is undesirable to transition to an arc discharge due to the intense heat generated, which causes deterioration of the cathode, as well as, localized hot spots which can lead to equipment failure. Therefore, there are several ways to avoid the transition to arcs, especially for atmospheric pressure discharges. The idea behind a *dielectric barrier discharge* is to place a dielectric material in between the two conducting electrodes to prevent the formation of arcs. A simple RF power supply with frequencies in the range of 0.5–500 kHz can be utilized to generate the discharge, so the simplicity of this configuration is part of the reason why this type of discharge has been extensively studied and used in a variety of applications, such as ozone generation, surface treatment, and pollutant decomposition,

Figure 3.7 Dielectric barrier discharge in atmospheric pressure air.

among others. Figure 3.7 shows a DBD in atmospheric air with a thin wire in the center of a ceramic tube which has its external surface wrapped in an aluminum film. A number of configurations have been tested, some of them with a dielectric material in contact with each electrode, while others include a single dielectric barrier attached to only one of the electrodes or in the center of the gap between the electrodes. Typical voltages used are in the $\pm 10\,\text{kV}$ range, with peak current below 0.1 A, and electron densities between 10^{20} and $10^{21}\,\text{m}^{-3}$.

3.2.3 Pulsed Discharges

Another method to avoid the formation of electric arcs in discharges of reasonable power is to provide high voltage pulses with a duration that is shorter than the time that it takes a streamer to reach from one electrode to the other. Since streamer propagation velocity is of the order of $10^6\,\text{m/s}$ and the gap is of the order of 1 cm, the duration of the pulse needs to be below 10^{-8} seconds, which means that more sophisticated (and expensive) nano-second pulse power supplies are needed to generate these discharges. Due to the fast electrons generated by these nonthermal discharges, they are very efficient in producing highly reactive radical species as well as energetic photons. They have been utilized in the past for control of pollutants, odor, contamination in water, biological tissue, and surface treatment, and also for ozone generation and volatile organic compound (VOC) decomposition. One of the easiest ways to achieve nano-second rise times is by means of a capacitive-storage pulse source, which stores charge in a capacitor that is discharged by a spark gap or magnetic switch (Huiskamp, 2020). Operating conditions depend on wire diameter, but measured values for applied voltage range between 15 and 40 kV, with electric fields between 9 and 23 kV/mm and streamer head velocities in the range between 1.48 and 2.14 ×10⁶ m/s (Wang and Namihira, 2020).

3.2.4 Gliding Arc

Gliding arc discharges have been gaining popularity in recent years because they are easy to implement and they provide good performance in methane partial oxidation (Kalra et al., 2005) and other plasma-assisted reforming applications (Petitpas et al., 2007). Figure 3.8 shows a picture of a functioning gliding arc discharge in atmospheric-pressure air, where the discharge is formed by applying a high electric potential difference between two electrodes shaped as diverging arcs (or diverging coils). The discharge forms at the closest point between the two electrodes, which is located right above the orifice observed at bottom plate. Gas (in this case air) flows from the orifice, dragging the arc formed between the two electrodes in the upward direction. As the arc elongates due to the diverging shape of the electrodes,

Figure 3.8 Gliding arc discharge.

it draws more power until the power supply can no longer maintain the arc, which vanishes. A new arc is formed at the narrowest gap between the electrodes, and the cycle is repeated again. An interesting aspect of gliding arc discharges that is described in detail in Mutaf-Yardimci et al. (2000) is that, depending on the operating conditions, they can display both thermal and nonthermal regimes. The authors found that nonequilibrium discharges occur at low current and high flow velocities, while thermal gliding arcs are attained as a quasiequilibrium discharge at high currents and low flow velocities, where the power per unit length remains almost constant through the arc evolution. The nonthermal gliding arc (or transition gliding discharge) is obtained at moderate values of current and flow velocity. The ionization mechanism changes from thermal ionization during the thermal arc phase to ionization due to high temperature electrons during the nonequilibrium phase. The authors experimentally observed that the transition from equilibrium to nonequilibrium phases can be identified by a change in voltage increase rate with respect to discharge-length growth.

3.2.5 Microwave-Induced Discharges

As the frequency of an AC voltage continues to be increased to values larger than that of radio frequencies, the range of the electromagnetic spectrum referred to as Microwaves is reached, which covers frequencies between 300 MHz and 300 GHz that have free-space wavelengths from 1 m to 100 mm, respectively. Since this portion of the spectrum includes the designated ranges of ultra high frequency, super high frequency, and extremely high frequency, most of these frequencies are reserved for communications, navigation, and military applications. However, a specified set of frequencies, such as 922 MHz, 2.45 GHz, and 5.8 GHz, among a few others, are allowed to be used for industrial applications. Metaxas and Meredith (1983) established that insulating materials can be heated by applying high-frequency electromagnetic waves. Microwave heating accounts for heat and moisture diffusion through the material. The power dissipated per unit volume in a nonmagnetic uniform material is given by:

$$P = 55.63 \times 10^{-12} f(\tau E)^2 \kappa'' \tag{3.12}$$

where f is the frequency, κ'' is the dielectric loss factor of the material, and τ is the transmission coefficient for incoming microwave (Brodie, 2012). An aspect unique to RF and microwave heating is that when the local diffusion rate is much smaller than the electromagnetic power dissipation rate, the local temperature increases rapidly and it might lead to thermal runaway.

For a treatment in depth of microwave heating, the reader is referred to Brodie et al. (2015).

3.3 Impedance Matching

As described in this chapter, depending on the type of application, non-thermal plasma discharges can be designed to operate with a wide range of input power, which covers from DC to RF (from kilohertz frequencies up to microwaves). In general, commercially available RF power supplies and amplifiers are designed to deliver power into a 50 Ω load. However, plasma reactors tend to have impedances that are very different than 50 Ω. Capacitively coupled plasma tends to behave as large capacitors in series with a small resistance, while inductively coupled plasma discharges behave as inductors. By matching the impedance at the output of the generator with the load impedance, maximum power transfer is allowed, which translates into efficient use of the power supply or amplifier. In addition, impedance matching minimizes the power reflected back to the power supply, which can damage the equipment. Several noninvasive methods have been proposed to measure the impedance of plasma discharges (Ilić, 1981; Zito et al., 2010).

The most common method for impedance matching is by means of utilizing a matching network. Maximum power transfer is obtained by matching the source (Z_s) and load (Z_p) impedances. The matching network is connected between the source (power supply or amplifier) and the load (plasma reactor), and it is designed to transform the output impedance of the source such that it is equal to the complex conjugate of the load impedance (Marovich, 2021; Zito et al., 2010). There are several approaches for the design of matching networks, some are more complex than others. One of the simplest and most widely used approaches is the L-network, shown in Figure 3.9. It is noted that in many implementations of the L-network, an additional tuning capacitor is added in series with the inductance. Defining the plasma impedance $Z_p = R_p + jX_p$, where R_p is the resistive part of the impedance and X_p is the reactive part, and j is the imaginary unit, and where R_s is source resistance, the components of the L-network can be computed as follows (COMSOL, 2012; Lieberman and Lichtenberg, 2005; Wang et al., 2019).

$$X_m = \sqrt{R_p R_s - R_p^2} - X_p \tag{3.13}$$

$$B_m = \sqrt{\frac{1}{R_p R_s} - \frac{1}{R_s^2}} \tag{3.14}$$

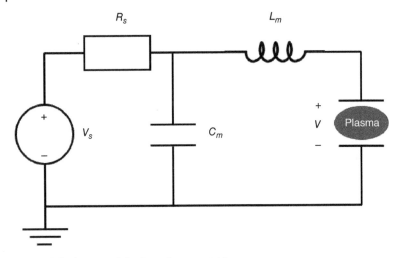

Figure 3.9 L-network for impedance matching.

Then, the matching network inductance and capacitance are calculated as:

$$L_m = \frac{X_m}{2\pi f} \tag{3.15}$$

$$C_m = \frac{B_m}{2\pi f} \tag{3.16}$$

where f is the frequency. The matching network intends to maximize the maximum power transfer coefficient:

$$\alpha = \frac{P_{plasma}}{P_{max}} \tag{3.17}$$

and power efficiency:

$$\eta = \frac{P_{plasma}}{P_{plasma} + P_s} \tag{3.18}$$

The impedance matching analysis can also be performed graphically using Smith charts (Chan and Harter, 2000). For cases of low power and when applications require frequencies below 1 MHz, impedance matching can also be performed using a transformer. This method is easy to implement and is less expensive than the impedance matching network, but it only matches the resistive portion of the load (Advanced-Energy-Industries, 2020).

3.4 Discharges in Liquids

Plasma discharges related to energy conversion have traditionally been applied in pure or combination of gases (e.g. plasma-enhanced methane steam reforming, tar breakdown, etc.) or in applications where plasma interacts with solid materials (e.g. plasma gasification, surface cleaning, plasma etching, etc.) However, a growing volume of research and development has focused on plasma discharges in liquids. Part of the reason why there are significantly less publications related to plasma discharges in liquids compared to gases is because there is still a lack of complete understanding of many aspects of these discharges. On the other hand, plasma discharges in liquids open up a completely new area of research which can enable a number of inventions and technological advances in processes related to energy and the conversion of carbonaceous materials.

3.4.1 Contact Glow Discharge Electrolysis

Contact glow discharge electrolysis (CGDE) also known as plasma electrolysis is a process discovered in the nineteenth century that has received recent attention due to its novel applications in the oil industry and in wastewater treatment. A sheath of plasma forms around an electrode immersed in liquid, producing an intense light, transferring heat to the fluid causing evaporation, and establishing a rotating cell motion of the fluid near its top surface. Figure 3.10 shows the setup to generate CGDE. An electrically conductive electrolyte solution is added to a metallic container that is connected to the positive electrode (anode) of the power supply. A cathode is located

Figure 3.10 Setup to produce contact glow discharge electrolysis.

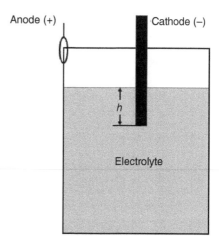

at the center of the container having the electrode immersed at a depth of around $h = 1$ cm inside the electrolyte. Common electrolytes are $NaHCO_3$, NaOH, or KOH at concentrations of 1–2% by weight, but the discharge can be operated with waste water, sea water, or other waste liquid streams. DC power is provided and the system initially operates under normal electrolysis, as seen in Figure 3.11. As the fluid near the cathode starts to heat up due to Joule heating, a convective current is formed which is partly due to natural convection but also due to EHD effects. Normal electrolysis is characterized by relatively low voltages and high current. A wide range of voltages can be utilized depending on the dimensions and characteristics of the electrode and tank as well as the electrolyte being used. The author's group performed tests using an ESAB ES-150 DC power supply which provided a maximum current of 150 A and a maximum voltage of 380 V (Sharma et al., 2013, 2014, Sharma and Diaz, 2015). The CGDE reactor used was obtained from Foret Plasma Labs, LLC (Foret, 2009). The temperature of the fluid near the cathode continues to increase until it is close to boiling. Since only a small section of the cathode is immersed in the electrolyte, current density is high and a thin film of vapor is formed around the cathode. At this point, transition from normal electrolysis occurs which can be identified by a drop in current and a simultaneous increase in voltage. A characteristic glow is then observed at the cathode as shown in Figure 3.12. The increase in voltage and reduction in current can be observed in Figure 3.13, where the voltage is seen to vary from around 100 to 380 V and the current varies from 125 A to

Figure 3.11 Normal electrolysis in water with 1% $NaHCO_3$.

Figure 3.12 Contact glow discharge electrolysis in water with 1% NaHCO$_3$.

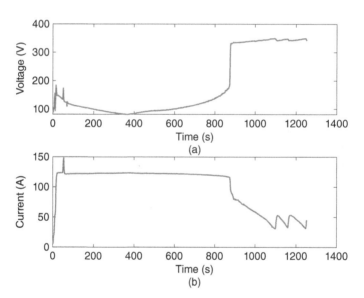

Figure 3.13 Transition from normal electrolysis to contact glow discharge electrolysis. (a) Voltage vs. time, (b) Current vs. time.

a value around 50 A. The sawtooth variation of the current (also observed in the voltage) after 1100 seconds occurs because as the liquid evaporates, the current continues to decay, but there is a point where the cathode might start to lose contact with the electrolyte. Before the discharge extinguishes, more electrolyte is added to the tank which increases the portion of the cathode immersed in the electrolyte allowing more current to flow. It is noted that if a large amount of electrolyte is added to the tank, then a large portion of the cathode becomes immersed, lowering the current density and transitioning back to normal electrolysis. With the type of electrolyte tested, steam and non-condensable gases were produced. Steam generation efficiency was 67% but it could be increased to 80% by adding a dielectric material such as gravel around the cathode to increase the region covered with CGDE. Non-condensable gases accounted for around 1–2% of the output gases by weight and were mainly composed of 45% of H_2, 33% of CO_2 and approximately 12% of O_2 with smaller fraction of other gases being present.

As mentioned above, the purpose of our tests were to generate steam; however, this processes can be used for a number of applications, including processing waste water from scrubbers in biomass gasification plants, UV treatment of liquid streams, and hydrogen production, among others. A review of this type of discharge and its applications can be found in Gupta (2015) and Yerokhin et al. (1999).

3.4.2 Plasma Electrolysis with AC Power

The large availability of biomass and the promulgation of stricter regulations for emissions of greenhouse gases and other pollutants from open burning, are fostering research in finding novel ways to develop value-added products from low-cost feedstocks. High-quality lignin and significantly higher yields of phenolic monomers can be obtained by depolymerization of the lignin produced utilizing plasma electrolysis of red oak immersed in γ-valerolactone (GVL) with sulfuric acid acting as the electrolyte (Lusi et al., 2021). Voltages in the range between 5 and 10 kV were used by the authors of the study with a frequency of 6 kHz in a 30 ml three-necked round-bottom flask. The same authors utilized plasma electrolysis of cellulose to achieve liquefaction of the cellulose obtaining 43% of levoglucosenone after 15 minutes of application of 6 kV AC power at a frequency of 6 kHz (Lusi et al., 2020).

3.4.3 Gliding Arc in Glycerol for Hydrogen Generation

Glycerol is an organic compound with the chemical formula $C_3H_8O_3$, which is a byproduct of several industrial processes including biodiesel production from biomass. Conventional steam reforming produces hydrogen by the reaction (Adeniyi, 2019):

$$C_3H_8O_3 + 3H_2O \rightarrow 7H_2 + 3CO_2 \tag{3.19}$$

However, formation of hydrogen can also be achieved under the presence of a plasma discharge. Gliding arcs were described in Section 3.2.4, where a discharge was formed between two diverging electrodes where the plasma was dragged in the upward direction by its interaction with a flow of a gas injected below the discharge. A similar discharge can be generated in liquids, as shown in Figure 3.14, where the buoyancy effect moves the arc upward without the need of injecting a gas or liquid. A gas film is formed around the arc, and due to buoyancy effects, it generates a bubble that grows until it separates from the discharge. The gas bubbles formed during the process are collected in a cylindrical section at the top of the reactor. The discharge's length continues to grow until the power supply can no longer increase power and the discharge vanishes. A new arc forms at the shortest gap between the electrodes and the process is repeated. Power used in this experiment ranged from 60 to 300 W, at 30 kV peak-to-peak, with a frequency of 30 kHz, and with a gap distance of 2 mm for tungsten electrodes. Analysis of the gases collected showed approximately 71% of H_2, 12% of CO_2, and 15% of CO (Plotnikov, 2019).

Figure 3.14 Gliding arc in glycerol.

Bibliography

A.G. Adeniyi. A review of steam reforming of glycerol. *Chemical Papers*, 73:2619–2635, 2019.

Advanced-Energy-Industries. Impedance matching. Technical Report 1-14, Advanced Energy Industries, Inc., 2020.

D.S. Antao, D.A. Staack, A. Fridman, and B. Farouk. Atmospheric pressure DC corona discharges: operating regimes and potential applications. *Plasma Sources Science and Technology*, 18(035016):1–11, 2009.

M.I. Boulos. Thermal plasma processing. *IEEE Transactions on Plasma Science*, 19(6):1078–1089, 1991.

G. Brodie. Application of microwave heating in agriculture and forestry related industries. In *The Development and Application of Microwave Heating*. Ed. W. Cao. InTech: Rijeka, Croatia, 45–78 2012.

G. Brodie, M.V. Jacob, and P. Farrell. *Microwave and Radio-Frequency Technologies in Agriculture*. De Gruyter Open, 2015.

C.M. Chan. Surface treatment of polypropylene by corona discharge and flame. In *Advanced Physicochemical Treatment Technologies, Polypropylene. Polymer Science and Technology Series*, Vol. 2. Ed. J. Karger-Kocsis. Springer, 1999. https://doi.org/10.1007/978-94-011-4421-6_109.

K.-C. Chan and A. Harter. Impedance matching and the smith chart - the fundamentals. *RF Design*, 52–66, 2000.

J. Chen and J.H. Davidson. Electron density and energy distributions in the positive dc corona: interpretation for corona-enhanced chemical reactions. *Plasma Chemistry and Plasma Processing*, 22:199–224, 2002.

D.F. Colas, A. Ferret, D.Z. Pai, D.A. Lacoste, and C.O. Laux. Ionic wind generation by a wire-cylinder-plate corona discharge in air at atmospheric pressure. *Journal of Applied Physics*, 108(10):103306, 2010.

COMSOL. Impedance matching. Technical Report 1-18, COMSOL Multiphysics, Model Manual 5.6, 2012.

C. Deron, P. Riviere, M.-Y. Perrin, and A. Soufiani. Coupled radiation, conduction, and joule heating in argon thermal plasmas. *Journal of Thermophysics and Heat Transfer*, 20(2):211–219, 2006.

Y. Duan, C. Huang, and Q. Yu. Low-temperature direct current glow discharges at atmospheric pressure. *IEEE Transactions on Plasma Science*, 33(2):328–329, 2005.

G.J. Dunn and T.W. Eagar. Calculation of electrical and thermal conductivity of metallurgical plasmas. Bullettin/Circular WRC 357, Welding Research Council, Inc., 1990.

T. Foret. US 2009/0206721 A1, System, method and apparatus for coupling a solid oxide high temperature electrolysis glow discharge cell to a plasma arc torch, 2009. http://www.plasmawhirl.com.

A. Fridman and L.A. Kennedy. *Plasma Physcis and Engineering*. Taylor & Francis Books, 2004.

C.K. Gilmore and S.R.H. Barrett. Electrohydrodynamic thrust density using positive corona-induced ionic winds for in-atmosphere propulsion. *Proceedings of the Royal Society A*, 471(20140912):1–24, 2015.

S.K.S. Gupta. Contact glow discharge electrolysis: its origin, plasma diagnostics and non-faradaic chemical effects. *Plasma Sources Science and Technology*, 24:063001, 2015.

K. Hattori, Y. Ishii, H. Tobari, A. Ando, and M. Inutake. Stabilization of dc atmospheric pressure glow discharge by a fast gas flow along the anode surface. *Thin Solid Films*, 506–507:440–443, 2006.

T. Huiskamp. Nanosecond pulsed streamer discharges Part I: Generation, source-plasma interaction and energy-efficiency optimization. *Sources Science and Technology*, 29(023002):1–47, 2020.

D.B. Ilić. Impedance measurement as a diagnostic for plasma reactors. *Review of Scientific Instruments*, 52(10):1542–1545, 1981.

C.S. Kalra, A.F. Gutsol, and A.A. Fridman. Gliding arc discharges as a source of intermediate plasma for methane partial oxidation. *IEEE Transactions on Plasma Science*, 33(1):32–41, 2005.

X. Li, C. Tang, and X. Dai. Study of atmospheric pressure abnormal glow discharge. *Plasma Science and Technology*, 10(2):185–188, 2008.

M.-R. Liao, H. Li, and W.-D. Xia. Approximate explicit analytic solution of the elenbaas-heller equation. *Journal of Applied Physics*, 120(063304): 063304-1–063304-6, 2016.

M.A. Lieberman and A.J. Lichtenberg. *Principles of Plasma Discharges and Materials Processing*. Wiley, 2nd ed., 2005.

P. Lippens. Chapter 3 - low-pressure cold plasma processing technology. In *Plasma Technologies for Textiles, Woodhead Publishing Series in Textiles*. Ed. R. Shishoo, pages 64–78. Woodhead Publishing, 2007. ISBN 978-1-84569-073-1. https://doi.org/10.1533/9781845692575.1.64.

C.-J. Liu, G.-H. Xu, and T. Wang. Non-thermal plasma approaches in CO_2 utilization. *Fuel Processing Technology*, 58:119–134, 1999.

A. Lusi, H. Radhakrishnan, H. Hu, H. Hu, and X. Bai. Plasma electrolysis of cellulose in polar aprotic solvents for production of levoglucosenone. *Green Chemistry*, 22:7871–7883, 2020.

A. Lusi, H. Radhakrishnan, H. Hu, and X. Bai. One-pot production of oxygenated monomers and selectively oxidized lignin from biomass based on plasma electrolysis. *Green Chemistry*, 23:9109–9125, 2021.

J. Marovich. *Practical Guide to Radio-Frequency Analysis and Design*. EETech Media, LLC, 2021. https://www.allaboutcircuits.com.

A.C. Metaxas and R.J. Meredith. *Industrial Microwave Heating*. Peter Peregrinous, 1983.

N. Monrolin, O. Praud, and F. Plouraboué. Electrohydrodynamic ionic wind, force field, and ionic mobility in a positive DC wire-to-cylinders corona discharge in air. *Physical Review Fluids*, 3(063701):1–20, 2018.

J. Mostaghimi, L. Pershin, and S. Yugeswaran. Heat transfer in DC and RF plasma torches. In *Handbook of Thermal Science and Engineering*. Ed. F. Kulacki, Springer, 1–76 2017.

O. Mutaf-Yardimci, A.V. Saveliev, A.A. Fridman, and L.A. Kennedy. Thermal and nonthermal regimes of gliding arc discharge in air flow. *Journal of Applied Physics*, 87(4):1632–1641, 2000.

G. Petitpas, J.-D. Rollier, A. Darmon, J. Gonzalez-Aguilar, R. Metkemeijer, and L. Fulcheri. A comparative study of non-thermal plasma assisted reforming technologies. *International Journal of Hydrogen Energy*, 32:2848–2867, 2007.

V. Plotnikov. *Development of high-voltage systems for direct and surface plasma treatment of liquids in sustainable energy*. PhD thesis, University of California - Merced, 2019.

J.R. Roth. *Industrial Plasma Engineering*, vol. 1. Institute of Physics Publishing, 1995.

S. Samal. Thermal plasma technology: the prospective future in material processing. *Journal of Cleaner Production*, 142:3131–3150, 2017.

T. Shao, R. Wang, C. Zhang, and P. Yan. Generation of a planar direct-current glow discharge in atmospheric pressure air using rod array electrode. *Scientific Reports*, 7(2672):1–7, 2017. https://doi.org/10.1038/s41598-017-03007-1.

T. Shao, R. Wang, C. Zhang, and P. Yan. Atmospheric-pressure pulsed discharges and plasmas: mechanism, characteristics and applications. *High Voltage*, 3(1):14–20, 2018.

N. Sharma and G. Diaz. Heat transfer and flow-pattern formation in a cylindrical cell with partially immersed heating element. *American Journal of Heat and Mass Transfer*, 2(1):31–41, 2015.

N. Sharma, G. Diaz, and E. Leal-Quiros. Evaluation of contact glow-discharge electrolysis as a viable method for steam generation. *Electrochimica Acta*, 108:330–336, 2013.

N. Sharma, A. Munoz-Hernandez, G. Diaz, and E. Leal-Quiros. Contact glow discharge electrolysis in the presence of solid organic waste. In *Poster at: XV Latin American Workshop on Plasma Physics*, San José, Costa Rica, 2014.

B.D. Shaw. Regular perturbation solution of the Elenbaas-Heller equation. *Journal of Applied Physics*, 99(034906):034906-1–034906-6, 2006.

D. Staack, B. Farouk1, A. Gutsol, and A. Fridman. Characterization of a dc atmospheric pressure normal glow discharge. *Plasma Sources Science and Technology*, 14:700–711, 2005.

Y. Tanaka. Synthesis of nano-size particles in thermal plasmas. In *Handbook of Thermal Science and Engineering*. Ed. F. Kulacki. Springer, 1–38 2017.

D. Wang and T. Namihira. Nanosecond pulsed streamer discharges: II. Physics, discharge characterization and plasma processing. *Plasma Sources Science and Technology*, 29(023001):1–25, 2020.

J. Wang, S. Dine, J.-P. Booth, and E.V. Johnson. Experimental demonstration of multifrequency impedance matching for tailored voltage waveform plasmas. *Journal of Vacuum Science and Technology A*, 37(2):1–13, 2019.

T.-Y. Wen and J.-L. Su. Corona discharge characteristics of cylindrical electrodes in a two-stage electrostatic precipitator. *Heliyon*, 6(e03334):1–6, 2020.

T. Yamamoto and M. Okubo. Nonthermal plasma technology. In *Advanced Physicochemical Treatment Technologies.*, Handbook of Environmental Engineering, vol. 5, Ed. L.K. Wang, Y.T. Hung, and N.K. Shammas, pages 135–293. Humana Press, 2007. https://doi.org/10.1007/978-1-59745-173-4_4.

A.L. Yerokhin, X. Nie, A. Leyland, A. Matthews, and S.J. Dowey. Plasma electrolysis for surface engineering. *Surface and Coatings Technology*, 122(2):73–93, 1999. ISSN 0257-8972.

C. Yuan, A.A. Kudryavtsev, and V.I. Demidov. *Introduction to the Kinetics of Glow Discharges*. IOP Concise Physics. Morgan & Claypool Publishers, 2018. ISBN 9781643270609. https://books.google.com/books?id= 24JqDwAAQBAJ.

J.C. Zito, D.P. Arnold, R.J. Durscher, and S. Roy. Investigation of impedance characteristics and power delivery for dielectric barrier discharge plasma actuators, aiaa 2010-964. In *48th AIAA Aerospace Sciences Meeting Including the New Horizons Forum and Aerospace Exposition*, pages 1–17. AIAA, January 2010.

4

Voltage-Enhanced Processing of Biomass

4.1 Biomass Gasification with Thermal Plasma

One approach for utilizing plasma for the means of gasifying coal, municipal solid waste, or biomass is by using energy-intensive thermal plasma torches which operate at high temperatures and produce a hydrogen-rich synthesis gas from the conversion of the feedstock. Although this application is relatively new (Leal-Quiros, 2004, Shie et al., 2008, Matveev et al., 2013), the topic has been attracting significant attention in recent decades. Van Oost et al. (2009) used a low flow rate DC plasma torch with water stabilized arc to analyze the thermal decomposition of crushed wood. Fractions of hydrogen and carbon monoxide of the order of 46% and 44%, respectively, were obtained with very low quantities of CO_2. Rutberg et al. (2011) determined efficiency values for plasma gasification of wood residues. They obtained net energies near 4.6 MJ/kg of wood processed with positive net energy ratios of 3.3 for electricity production only, and of 7.9 for combined heat and electricity generation. Very recently, Kuo et al. (2021) utilized Aspen Plus$^{©}$ to model a DC plasma torch performing an energy and exergetic analysis as well as the life cycle analysis of greenhouse gas emissions for plasma gasification of microalgal biomass. They found that torrefaction of the microalgae improved syngas and hydrogen yield by 36.7% and 33.86%, respectively. Comprehensive reviews of this topic can be found in Fabry et al. (2013) and Sanlisoy and Carpinlioglu (2017).

4.1.1 Plasma Parameters

Thermal plasma torches can operate with a variety of working gases that are ionized and put in contact with biomass. The most common gases for the generation of plasma are air, nitrogen, argon, helium, and hydrogen

Voltage-Enhanced Processing of Biomass and Biochar, First Edition. Gerardo Diaz.
© 2022 John Wiley & Sons Ltd. Published 2022 by John Wiley & Sons Ltd.

(Venkatramani, 1995); however, other gases such as steam or methane can also be used depending on the application. Inert atmospheres are usually required in processes such as arc welding or for plasma cutters, thus, noble gases such as argon or helium tend to be used, with argon being preferred due to the higher cost of helium. On the contrary, reacting atmospheres are preferred in biomass processing, so air or nitrogen are commonly used, in addition to the lower cost of these gases. Nonetheless, certain plasma torches require small fractions of argon that is added in order to reduce ignition voltage of the discharge in the reactor (Komarzyniec and Aftyka, 2020).

Results obtained at the Sustainable Plasma Gasification Laboratory at the University of California, Merced using a two-stage plasma gasification system from Foret (2009) are presented in Sections 4.1.2 and 4.1.3. The first stage used a 1% solution of sodium bicarbonate in water and operates in glow discharge mode with the purpose of generating steam (Sharma et al., 2013a,b). A mixture of steam and a small fraction of non-condensable gases generated in the first stage is utilized as gasifying agent in a second stage which is composed of an ArcWhirl® reactor where a DC thermal plasma discharge is generated and put in contact with biomass to perform a steam plasma gasification process (Diaz et al., 2015). The reactor operates close to ambient pressure avoiding the large costs of high-pressure equipment commonly used in water-gas shift reactors utilized to increase hydrogen output.

The plasma gasification process is subject to variability due to several factors, such as fluctuations in biomass flow rate due to changes of size and composition of the feedstock, nonuniform expansion of gases inside the reactor, variability of the rate of cathode consumption, nonuniform residence times, and changes of temperature inside the reactor that vary the rates of thermochemical decomposition and syngas production. In addition, there are fluctuations in arc voltage and current, radiation heat loss, acoustic pressure effects (Brilhac et al., 1995, Singh et al., 2000), and the undulation of three-phase rectified power supply (Tu et al., 2008).

Thermal plasmas from DC torches have energy densities from 10^6 to 10^7 J/m^3, where high collision frequencies lead to local thermodynamic equilibrium, for which the kinetic energy of the species involved, i.e. electrons, ions, and neutral gas, is characterized by a single temperature. Temperature measured with a thermocouple immersed in the biomass (so it does not represent the plasma temperature), showed values in the range between 800 and 1000 °C, while measurements of the plasma temperature with an optical pyrometer showed temperatures at the discharge orifice around 3000 °C for steam plasma and 2200 °C for air plasma. Input power ranged from 10 to 30 kW, for steam flow rate between 10 and 12 kg/h, and biomass flow rate in the range from 3 to 5 kg/h.

4.1.2 Syngas Composition

Steam plasma gasification tests performed for seven different biomass types are shown in this section, where ultimate and proximate analyses results for the different biomass types are shown in Tables 4.1 and 4.2, respectively. A schematic of the two-stage plasma gasification system and sampling components is shown in Figure 4.1, where gas sample bottle GS1 is used as a sacrificial sample to remove contaminants from the line before syngas samples are obtained using bottles GS2 and GS3. Four measurements of syngas composition were made for each type of biomass, where gas compositions were determined using gas chromatography.

Table 4.1 Ultimate analysis (weight fractions) and LHV of biomass analyzed in this work.

Biomass	C (%)	H (%)	O (%)	N (%)	LHV (kJ/kg)
Almond shells	44.96	5.31	38.26	1.37	18,059.0
Hard wood shaving	48.41	6.28	41.1	0.13	18,854.5
PB&MDFB[a)]	4.98	6.16	41.41	3.53	19,289.5
Peach pits	52.52	6.18	39.74	0.38	20,754.9
Almond hulls	44.31	5.64	40.13	1.06	18,040.5
Pomace	52.74	6.23	33.48	2.14	21,883.0
Coffee ground	56.13	7.16	27.03	2.53	23,771.7

a) PB&MDFB: Particle Board and Medium Density Fiber Board.

Table 4.2 Proximate analysis of biomass analyzed (weight fractions).

Biomass	Ash (%)	Fixed carbon (%)	Moisture (%)	Sulfur (%)	Volatile matter (%)
Almond shells	10.0	17.1	5.66	0.045	67.2
Hard wood shaving	4.03	7.53	5.72	<0.01	7.53
PB&MDFB	0.87	14.7	6.32	0.027	78.1
Peach pits	0.72	11.9	36.73	0.46	50.7
Almond hulls	8.86	18.9	8.01	<0.01	64.3
Pomace	5.29	18.5	8.97	0.12	67.2
Coffee ground	1.14	7.07	54.69	0.08	37.1

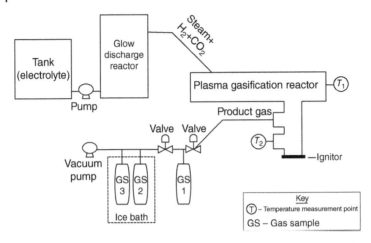

Figure 4.1 Synthesis gas measurements during steam plasma gasification tests.

Table 4.3 Synthesis gas composition for different types of biomass (molar basis).

Biomass	CH_4	C_2H_6	C_3H_8	C_4H_{10}	H_2	O_2	N_2	CO	CO_2	LHV[a]
Almond shells	2.4	0.0	0.0	0.0	73.7	0.2	1.9	5.1	16.7	9.5
Hard wood shaving	3.9	0.0	0.0	0.0	55.9	1.2	4.7	12.3	22.0	9.0
PB&MDFB	4.7	0.0	0.2	0.1	37.3	1.0	4.5	24.5	27.5	9.2
Peach pits	4.6	0.0	0.6	0.2	56.9	0.8	2.4	18.7	15.7	10.9
Almond hulls	1.8	0.0	0.0	0.1	52.2	1.2	4.4	11.8	28.4	8.0
Pomace	2.7	0.0	0.0	0.1	56.9	0.6	2.4	13.5	23.6	9.1
Coffee ground	2.5	0.0	0.0	0.0	77.0	0.2	1.3	4.1	14.9	9.7

a) LHV (MJ/Nm^3).

The average values of the product gas compositions for different species are shown in Table 4.3, where the ranges of operating conditions are shown in Table 4.4. The product gas from peach pits had the highest LHV at $10.9\,MJ/Nm^3$ possibly due to the high content of carbon monoxide (18.8%) and low content of carbon dioxide (15.7%) present in the samples. Almond hulls produced the gas with the lowest LHV at $8.0\,MJ/Nm^3$ which contained a low fraction of CO (11.8%) and CH_4 (1.8%) and the largest fraction of CO_2 of all the product gas tested (28.4%). These results show that steam plasma gasification generated very low quantities of higher hydrocarbons such

Table 4.4 Range of operating conditions.

Parameter	Range	units
Steam flow rate	10–12	kg/h
Power input (glow discharge)	11–12	kW
Power input (DC arc)	20–25	kW

as ethane, propane, butane, pentane, and hexane with fractions levels no larger than 0.6%. This can be explained due to the high chemical activity of plasma due to the presence of active species such as radicals, charged particles, electrons, ions, and excited atoms and molecules, which are efficient in breaking the chemical bonds in biomass and tars to generate small molecules such as H_2, CO, CO_2, and CH_4. These product gas compositions compare favorably against the composition obtained from a conventional down draft gasifier operating with Douglas-fir wood, which generated: 17.51% H_2, 10.98% CO_2, 20.01% CO, 1.98% CH_4, 47.47% N_2, and 1.95% O_2 (Emmerson and Diaz, 2010).

It is noted that the fractions of hydrogen are quite large, with most percentage values ranging from the mid-fifties to the seventies. This is explained due to the use of steam plasma; however, this comes with the cost of producing steam, which in these tests utilized between 11 and 12 kW to operate the glow discharge reactor. This additional cost can be reduced by using air instead of steam to generate the plasma at the torch; however, the fraction of hydrogen obtained in the product gas will be lower.

4.1.3 Energy Balance

A number of researchers have analyzed the process of plasma gasification by exploring the effects of different parameter values on the performance of experimental apparatus. However, there is a growing number of computational models being developed to analyze plasma torches and the gasification of carbonaceous materials. Detailed models of thermal plasma discharges (Trelles et al., 2009) provide an insight of the transport phenomena occurring in plasma torches. Other modeling approaches provide a description of an entire plasma gasification plant (Kuo et al., 2021). Due to the LTE condition, simplified thermodynamic models can be generated to obtain a perspective of the product gas compositions generated from plasma gasification of biomass (Diaz et al., 2014, Zainal et al., 2001, Mountouris et al., 2006). The following system of chemical reactions describes the main conversions during steam gasification of biomass.

$$C + CO_2 \leftrightarrow 2CO \quad \text{(Boudouard equilibrium)} \tag{4.1}$$

$$C + 2H_2 \leftrightarrow CH_4 \quad \text{(hydrogenating gasification)} \tag{4.2}$$

$$C + H_2O \leftrightarrow CO + H_2 \quad \text{(heterogeneous water}$$
$$\text{gas shift reaction)} \tag{4.3}$$

$$CH_4 + H_2O \leftrightarrow CO + 3H_2 \quad \text{(methane decomposition)} \tag{4.4}$$

$$CO + H_2O \leftrightarrow CO_2 + H_2 \quad \text{(water gas shift reaction)} \tag{4.5}$$

Low temperature or low-steam content promotes the formation of soot so the first three equations become relevant. At higher fractions of steam addition and high temperature, carbon (soot) is not formed so that the last two equations become dominant (Schuster et al., 2001). A global gasification reaction can be written as:

$$aC + bH + cO + dH_2O + eH_2 + fCO_2$$
$$= n_1H_2 + n_2CO + n_3CO_2 + n_4H_2O + n_5CH_4 + n_6C \tag{4.6}$$

where the coefficients a, b, c, d, e, and f are obtained from the ultimate analysis and the output from the steam generation stage, and coefficients n_1–n_6 are unknowns. Performing a molar balance for each species and adding three equilibrium constants from the methane decomposition, primary water gas, and water gas shift reactions, i.e.

$$k_1 = \frac{y_{CO} y_{H_2}^3}{y_{CH_4} y_{H_2O}}; \quad k_2 = \frac{y_{CO_2} y_{H_2}}{y_{CO} y_{H_2O}}; \quad k_3 = \frac{y_{CO} y_{H_2}}{y_{H_2O}} \tag{4.7}$$

the system of equations is solved, where the equilibrium constants k_1, k_2, and k_3 can be obtained from $\ln[k_i(T)] = \frac{\Delta G_i^\circ}{RT}$ for $i = 1, 2, 3$.

The experimental results of product gas composition and power consumption are compared against simulations with the thermodynamic model as observed in Figure 4.2 and Table 4.5 for grape pomace and almond hulls, respectively, where reasonable agreement is observed. The system

Table 4.5 Comparison of power consumption and product gas generation for almond hulls.

	Model	Experimental
Power consumption for steam (kW)	8.7	10
Power consumption arc (kW)	16.7	15–20
Flow rate of dry product gas (Nm³/h)	12.5	10

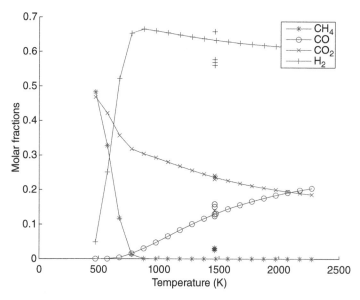

Figure 4.2 Comparison of thermodynamic model and experimental measurements for pomace biomass (Diaz et al., 2015).

was not optimized for waste energy recovery and thus, the ratio of the cold gas LHV to the biomass LHV was less than unity. However, several authors have demonstrated magnitudes between 3 and 5 for the ratio of calorific value of produced syngas with respect to the energy spent for its production (Van Oost et al., 2009), and net-energy ratios of 4.6–4.8 when both electricity and heat are considered (Rutberg et al., 2011).

4.1.4 Temperature Decay in Plasma/Biomass Discharge

The temperature of a plasma discharge decreases rapidly with distance from the nozzle, so in order to maximize its effects in the conversion, it is important to ensure that biomass is in direct contact with the discharge. As charged particles recombine and reactive species undergo reactions, the shorter the exposure (resident time) of the biomass particles to the discharge, the weaker the effect of the plasma on the gasification process. Figure 4.3 depicts an apparatus built to determine the temperature decay of a plasma discharge that is in contact with a stream of biomass such as hardwood shavings, sawdust, or coffee waste. Biomass is uniformly fed downward at the top of the "T" and gets in contact with an air-plasma discharge introduced from the right. As seen in the figure, the distance between the inlets of the plasma and biomass is approximately 8 cm.

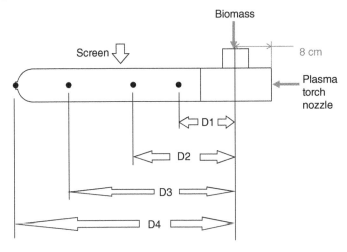

Figure 4.3 Apparatus to measure temperature distribution on a plasma discharge with biomass.

Plasma and biomass are combined at the "T" and the ionized mix exits to the left, where a metal cylindrical screen is fitted with thermocouples at increasing distances from the biomass inlet. The distances from the thermocouples to the biomass inlet correspond to: $D1$ (for TC 2) = 27.5 cm, $D2$ (for TC 3) = 37.5 cm, $D3$ (for TC 4) = 49.5 cm, $D4$ (for TC 5) = 73.5 cm, where thermocouple TC 1 measures the ambient temperature. The voltage supplied corresponds to 260 V and the current is 76 A. Measurement of the air velocity at the suction of the blower was 13.49 m/s with an inlet diameter of 9.5 cm, and a biomass flow rate of approximately 120 g/min. Figure 4.4 shows the temperature decay of an air plasma discharge combined with hardwood shavings with respect to the distance from the biomass inlet during a start-up and shut-down process. It is seen that at 57 seconds from the start of the discharge, the temperature for TC 2 located at 27.5 cm from the biomass inlet has reached 1000 °C. The temperature at this point is at steady state and it stays constant until the discharge is turned off at 73 seconds. The temperature for TC 3 located at a distance of 37.5 cm also approaches steady state. At discharge shut-down time, it is seen that the temperature is at 918 °C, which is 82 °C lower than TC 2. The maximum temperature for TC 5 located at a distance of 73.5 cm corresponds to 665 °C, which shows that in 46 cm the discharge temperature has dropped by 335 °C. Once the discharged is turned off at 73 seconds, all temperatures decay rapidly since the plasma discharge is no longer present.

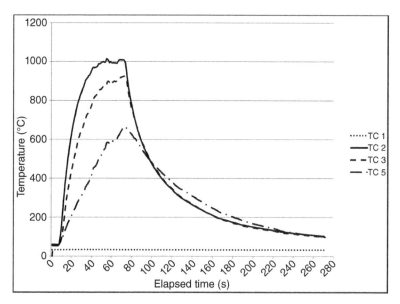

Figure 4.4 Temperature decay of plasma discharge with biomass with respect to distance.

4.2 Dielectric Breakdown of Biomass

Biomass is a poor electrical conductor and has been used as insulator for many applications. However, when exposed to high electric fields, and especially when its moisture content is high, a current can be circulated through this material, which increases its temperature due to Joule heating. As the temperature of the material increases and the moisture is evaporated, thermochemical conversion of the biomass occurs as cellulose and hemicellulose decompose and volatiles are released in a voltage-enhanced pyrolysis process. The carbonized material has a higher electrical conductivity which facilitates the flow of more current. This positive feedback loop can lead to the so called *thermal runaway*, where a sudden drastic increase in temperature occurs because the heat produced by the flow of current is larger than the rate of heat dissipation from the material, leading to biomass being converted to a material with properties similar to biochar with a high carbon content.

4.2.1 Biomass-in-the-Loop

Biomass gasification with thermal plasma torches is an energy intensive process where plasma generated by the DC torch provides the energy for the biomass conversion to product gases and biochar. This means that the carbonaceous material is exposed to the heat and reactive atmosphere generated by the plasma discharge. Since the voltages used by plasma torches are relatively low (of the order of 100–300 V), no electrical breakdown occurs so no current flows through the material during the process. A different approach where biomass becomes part of the electric circuit (biomass-in-the-loop) occurs when the voltage is increased beyond the point of electrical breakdown of the material, so that current starts to flow through it and its surface. As the material carbonizes, its electrical resistance decreases, making it easier for current to flow through it. As an arc forms between the high-voltage electrode and the biomass, the current flowing through the biomass generates Joule heating which increases the temperature of the material providing a different form of heating mechanism compared to the conventional DC plasma torches. Figure 4.5 shows a high-voltage gouging electrode (10 kV DC) in proximity to almond hulls forming a discharge between the two materials. Several high-voltage electrodes can be used to increase the power delivered to the biomass and to increase the points of contact between plasma and carbonaceous

Figure 4.5 High voltage (10 kV DC) discharge in contact with almond hulls.

material. Higher moisture contents in the biomass reduce the electric resistance and thus, the technique can still be applied for conditions where conventional autothermal thermochemical conversion processes cannot be utilized. Although this is an efficient method to deliver energy for biomass conversion, it is not commonly utilized as compared to thermal plasma torches.

4.3 Biomass Gasification with Nonthermal Plasma

As mentioned in Sections 4.1 and 4.2, the intense heat delivered by thermal plasma discharges is the main source of energy used for plasma-enhanced thermochemical decomposition of biomass. However, a different approach is to utilize cold or nonthermal plasma to enhance conventional processes by providing a highly reactive environment that facilitates the conversion of the biomass material and product gas. Neutral particles and ions remain close to ambient temperature in nonthermal plasmas, but the electrons have energies of several electron volts. These energized electrons can be effectively used to break down pollutant molecules and thus, provide an alternative method for the product-gas cleanup process. Conventional gasification systems used for power generation require several cleaning stages before the product gas can be injected in generator sets. Failure to clean the product gas results in pipe clogging and short operational life of the internal combustion engine. Concentration of tar in the product gas is recommended to be below $100 \, mg/Nm^3$ (Baratieri et al., 2009). As tars condense on pipe walls, a progressive reduction of power is also observed in the system. The gas cleanup stages not only require costly maintenance of the equipment but they also generate wastewater and oil contaminated with tars at the scrubbers. An appropriate form of liquid waste disposal is therefore required, which adds to the cost of the overall biomass conversion process. One alternative to reduce gas clean up is to use nonthermal plasma to break down tars in the product gas as discussed in Section 4.3.1.

4.3.1 Tar Breakdown

Tars are a mixture of organic compounds composed of heavy hydrocarbons, which condense at medium to high temperatures. Tars constitute one of the main issues that prevent wide-spread use of gasification technology for power generation. Gas cleanup stages usually involve cyclones, thermal cracking, catalytic cracking, water scrubber, oil scrubber, filters, gas coolers and reheaters, before the syngas can be injected into the engine. The gas cleanup constitutes a critical part of a biomass gasification plant which requires significant maintenance effort and cost to operate.

Nonthermal plasma discharges use a fraction of the energy needed by their thermal plasma counterparts; however, the energy carried by the electrons is significant, allowing these discharges to effectively break the bonds of large organic molecules reducing them down to small molecules such as H_2, CO, CH_4, among others. This effect can be utilized to decompose tar molecules in the product gas to increase the fractions of H_2 and CO produced.

4.3.1.1 Model Tars

Biomass thermochemical decomposition by gasification is a complex process that generates mainly product gas, tars, and biochar. The product gas includes a fraction of syngas (H_2 and CO) as well as other gases such as CO_2, N_2, CH_4, O_2 and some larger molecules such as ethane and butane. On the other hand, a very large variety of heavy hydrocarbons classified as either gravimetric or GC-detectable tars are produced. Therefore, a first step to analyze the effect of nonthermal plasma in the form of a dielectric barrier discharge (DBD) in breaking tars is to utilize a model tar that can be easily measured before and after the plasma reactor. Figure 4.6 shows the schematic of a DBD reactor, where Figure 4.6a shows the side view of the reactor with a thin (1.6 mm diameter) high-voltage electrode located in the middle of a cylindrical dielectric (ceramic) tube of 19 mm of internal diameter. The ceramic tube is surrounded by a metallic cylinder that acts as the ground electrode. Figure 4.6b shows the front of the reactor where it is observed that the plasma discharge fills the entire volume. Results were obtained for toluene decomposition at a flow rate of 500 ml/min of a calibrated mixture of toluene and air (20.9% oxygen in nitrogen) was passed through the reactor, where Table 4.6 shows the concentration of toluene in the gas.

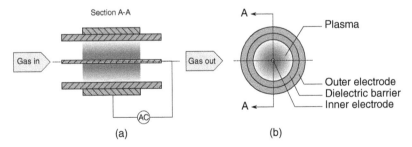

Figure 4.6 Schematic of a DBD tar breakdown reactor. (a) Side view, (b) front view.

Table 4.6 Toluene ppm before plasma.

Sample	1	2
Toluene (ppm)	119	120.7

Table 4.7 Toluene ppm after DBD plasma.

Sample	1 (ppm)	2 (ppm)
Toluene (ppm)	45.3	39.2
Toluene decomposition (%)	62.1	67.5

AC power was supplied to the reactor at 60 W, 15 kHz, and 30 kV peak-to-peak, where gas samples were collected with a 6 ml syringe and directly injected into a gas chromatograph. Table 4.7 shows the measurements of toluene concentration after one pass through the DBD reactor, where tests were performed at different specific energy inputs (Gomez et al., 2021).

It is observed that DBD was effective in reducing the concentration of the model tar in the gas mixture. Toluene concentration after plasma treatment was 45.3 and 39.2 ppm. These concentrations translate into reductions of 62.1% and 67.5% with respect to the inlet toluene concentration for samples 1 and 2, respectively. Increase in residence time or multiple passes through the reactor can decrease tar concentrations even more. It is noted that a number of researchers have obtained promising results for toluene, benzene, phenol, and naphthalene removal with pulsed corona or DBD discharges. A review of the subject can be found in Rueda and Helsen (2020) and Saleem et al. (2020).

4.3.1.2 Product Gas from Gasification/Pyrolysis

The thermochemical conversion of biomass through pyrolysis or gasification produces a number of tar species that vary depending on gasifier type, feedstock, and operating conditions. Authors have classified tars based on processes, such as primary, secondary, and tertiary, or by molecular weight from Class 1 to 5, being described by GC-undetectable, heterocyclic aromatics, light aromatics, light PAH compounds, and heavy PAH compounds (Rueda and Helsen, 2020). Tars have also been described by maturation schemes based on temperature, such as shown in Table 4.8 (Elliott, 1988, Milne et al., 1998).

Table 4.8 Tar maturation.

Temperature (°C)	Tar type
400	Mixed oxygenates
500	Phenolic ethers
600	Alkyl phenolics
700	Heterocyclic ethers
800	PAH
900	Larger PAH

Sources: Elliott (1987) and Milne et al. (1998).

Due to this complexity, most authors have used nonthermal plasma discharges to breakdown model tars, where conversion is analyzed for a few types of species. However, a few works have analyzed the use of nonequilibrium plasmas to breakdown tars formed in actual gasification processes. Craven et al. (2020) tested an integrated and nonthermal plasma-catalysis system to produce cleaner syngas from cellulose. Their system showed reduction in the concentration of tars by 88% and an increase in hydrogen production of 92%. In a different study, Blanquet et al. (2019) tested a laboratory-scale reactor using pyrolysis with nonthermal plasma and a nickel-based catalyst. Results showed a reduction in hydrocarbon tar content of 21% for the pyrolysis with plasma case, and 64% for the pyrolysis with plasma-catalyst case compared to pyrolysis alone. Figure 4.7 shows a DBD reactor constructed at the University of California – Merced for decomposition of tars from conventional gasification of biomass. The power supply provides up to 300 W at 15 kHz with 30 kV peak-to-peak. Reductions of 90% in tar concentration and more than twice the hydrogen content were obtained from tests at 150 W of DBD plasma discharge applied to the product gas from a downdraft gasifier (Indrawan et al., 2021).

4.3.2 Circuit Configuration

The literature related to high-voltage AC oscillators is quite large. One of the main issues preventing widespread utilization of nonthermal plasma is related to the cost of high-voltage power supplies. Most commercially available units cost between thousands to tens of thousands of dollars. In many cases, AC oscillators that can provide kilohertz frequencies with voltage ranges between 10 and 30 kV peak-to-peak are suitable for a variety of energy-related applications, including voltage-enhanced processing of

Figure 4.7 DBD reactor for tar decomposition.

carbonaceous materials as well as treatment of gases and liquids. Circuits with these specifications can be achieved at a cost of the order of only hundreds of dollars.

Soft switching is a technique for controlling an electrical switch with minimized switching losses. A zero voltage switching (ZVS) was utilized to develop a low-cost high-voltage power supply (Plotnikov, 2019) that was used to obtain some of the nonthermal plasma results shown in Sections 4.3.1.1 and 4.3.1.2. The technique is described in detail by Edry and Ben-Yaakov Edry and Ben-Yaakov (1993). The gate drive circuitry of the MOSFET can be connected to an LC tank by means of reference diodes. Figure 4.8 shows a simplified schematic diagram of the inverter, where the load (R_{load}) is powered by a high-voltage step-up transformer that consists of two primary windings L_1 and L_2 (close to $5\,\mu H$) and the secondary winding L_3 (close to 1 H). The primary winding establishes an LC tank with the capacitor C_1. The switching of power input (V_1) is performed by the two MOSFETs Q_1 and Q_2 that are referenced to the LC tank through Zener diodes D_2 and D_3. Protection of the DC power input V_1 includes diode D_1. In addition, the choke L_4 is added in order to minimize high-current transient

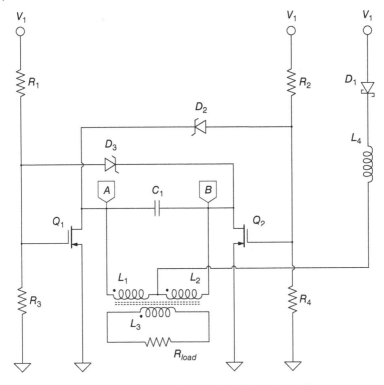

Figure 4.8 Simplified schematic diagram of a high-voltage AC generator.

during the starting operation of the power supply. Frequency control can be carried out by adjustment of capacitance in the LC tank.

4.3.3 Scaling Up of the Technology

Input power for thermal plasma torches ranges from around 10 kW to roughly 10 MW, which makes them suitable for a wide range of applications from small to large scale. This technology is quite effective in treating hazardous waste. However, the investment and operational costs for processing municipal solid waste based on charging tipping fees is less feasible in most areas of the world. The requirement of access to large amounts of power also makes this technology hard to implement in remote locations to process forest or agricultural waste. Nonetheless, there are a number of applications where DC thermal torches are suitable for processing waste.

On the other hand, atmospheric pressure nonthermal plasma discharges are relatively new and there are tremendous opportunities for applications dealing with processing of carbonaceous materials. There is a growing volume of research analyzing the effects of nonthermal plasma and nonequilibrium plasma in combination with catalysts. Some recent reviews of this subject can be found in Bogaerts et al. (2020) and Mehta et al. (2019). This technology is quite suitable for tar decomposition of the product gas from pyrolysis and gasification plants from small to medium scale. The technology is also suitable for treatment of liquid waste generated in biomass processing plants. On the other hand, implementation of nonthermal plasma directly at the gasification or pyrolysis reactor is still at the research stage.

Bibliography

M. Baratieri, P. Baggio, B. Bosio, M. Grigiante, and G. Longo. The use of biomass syngas in IC engines and CCGT plants: a comparative analysis. *Applied Thermal Engineering*, 29:3309–3318, 2009.

E. Blanquet, M.A. Nahil, and P.T. Williams. Enhanced hydrogen-rich gas production from waste biomass using pyrolysis with non-thermal plasma-catalysis. *Catalysis Today*, 337:216–224, 2019. ISSN 0920-5861. https://doi.org/10.1016/j.cattod.2019.02.033. Frontiers in Plasma Catalysis (ISPCEM 2018).

A. Bogaerts, X. Tu, J.C. Whitehead, G. Centi, L. Lefferts, O. Guaitella, F. Azzolina-Jury, H.-H. Kim, A.B. Murphy, W.F. Schneider, T. Nozaki, J.C. Hicks, A. Rousseau, F. Thevenet, A. Khacef, and M. Carreon. The 2020 plasma catalysis roadmap. *Journal of Physics D: Applied Physics*, 53(44):443001, 2020. https://doi.org/10.1088/1361-6463/ab9048.

J.F. Brilhac, B. Pateyron, J.F. Coudert, P. Fauchais, and A. Bouvier. Study of the dynamic and static behavior of de vortex plasma torches: Part II: Well-tye cathode. *Plasma Chemistry and Plasma Processing*, 15(2):257–277, 1995.

M. Craven, Y. Wang, H. Yang, C. Wu, and X. Tu. Integrated gasification and non-thermal plasma-catalysis system for cleaner syngas production from cellulose. *IOP SciNotes*, 1(2):024001, 2020. https://doi.org/10.1088/2633-1357/aba7f6.

G. Diaz, E. Leal-Quiros, R.A. Smith, J. Elliott, and D. Unruh. Syngas generation from organic waste with plasma steam reforming. *IOP Journal of Physics: Conference Series*, 511(012081):1–6, 2014. https://doi.org/10.1088/1742-6596/511/1/012081.

G. Diaz, N. Sharma, E. Leal-Quiros, and A. Munoz-Hernandez. Enhanced hydrogen production using steam plasma processing of biomass: experimental apparatus and procedure. *International Journal of Hydrogen Energy*, 40:2091–2098, 2015.

D. Edry and S. Ben-Yaakov. Capacitive-loaded push-pull parallel-resonant converter. *IEEE Transactions on Aerospace and Electronic Systems*, 29(4):1287–1296, 1993.

D.C. Elliott. Relation of reaction time and temperature to chemical composition of pyrolysis oils. In *Pyrolysis Oils from Biomass, ACS Symposium Series 376*. Ed. E.J. Soltes and T.A. Milne. ACS Publications: Denver, CO, 55–65 1988.

V.M. Emmerson and G. Diaz. Experimental characterization of a small-scale downdraft gasifier for biomass waste. In *proceedings of ASME IMECE 2010, Paper # IMECE2010-37392*, pages 1–5, Vancouver, Canada, November 2010.

F. Fabry, C. Rehmet, V. Rohani, and L. Fulcheri. Waste gasification by thermal plasma: a review. *Waste and Biomass Valorization*, 4(3):421–439, 2013.

T. Foret. US 2009/0206721 A1, System, method and apparatus for coupling a solid oxide high temperature electrolysis glow discharge cell to a plasma arc torch, 2009. http://www.plasmawhirl.com.

H. Gomez, V. Plotnikov, and G. Diaz. A fundamental parametric study and reaction kinetics of toluene decomposition using non thermal plasma. In *5-6th Thermal and Fluids Engineering Conference (TFEC)*, (Virtual presentation), 2021. American Society of Thermal Fluids Engineers. https://doi.org/10.1615/TFEC2021.fnd.032461.

N. Indrawan, V. Plotnikov, S. Thapa, G. Diaz, A. Kumar, and R.L. Huhnke. Performance evaluation of non-thermal plasma for syngas reforming. *in Preparation*, 2021.

G. Komarzyniec and M. Aftyka. Operating problems of arc plasma reactors powered by AC/DC/AC converters. *Applied Sciences*, 10(3295):1–14, 2020.

P.-C. Kuo, B. Illathukandy, W. Wu, and J.-S. Chang. Energy, exergy, and environmental analyses of renewable hydrogen production through plasma gasification of microalgal biomass. *Energy*, 223:120025, 2021. ISSN 0360-5442. https://doi.org/10.1016/j.energy.2021.120025. https://www.sciencedirect.com/science/article/pii/S0360544221002747.

E. Leal-Quiros. Plasma processing of municipal solid waste. *Brazilian Journal of Physics*, 34(4B):1587–1593, 2004.

I.B. Matveev, N.V. Washcilenko, S.I. Serbin, and N.A. Goncharova. Integrated plasma coal gasification power plant. *IEEE Transactions on Plasma Science*, 41(12):3195–3199, 2013.

P. Mehta, P. Barboun, D.B. Go, J.C. Hicks, and W.F. Schneider. Catalysis enabled by plasma activation of strong chemical bonds: a review. *ACS Energy Letters*, 4:1115–1133, 2019.

T.A. Milne, R.J. Evans, and N. Abatzoglou. Biomass gasifier tars: their nature, formation, and conversion. Technical Report NREL/TP-570-25357, National Renewable Energy Laboratory, November 1998.

A. Mountouris, E. Voutsas, and D. Tassios. Solid waste plasma gasification: equilibrium model development and exergy analysis. *Energy Conversion and Management*, 47(13–14):1723–1737, 2006.

V. Plotnikov. *Development of high-voltage systems for direct and surface plasma treatment of liquids in sustainable energy.* PhD thesis, University of California - Merced, 2019.

Y.G. Rueda and L. Helsen. The role of plasma in syngas tar cracking. *Biomass Conversion and Biorefinery*, 10:857–871, 2020. https://doi.org/10.1007/s13399-019-00461-x.

Ph.G. Rutberg, A.N. Bratsev, V.A. Kuznetsov, V.E. Popov, A.A. Ufimtsev, and S.V. Shtengel. On efficiency of plasma gasification of wood residues. *Biomass and Bioenergy*, 35:495–504, 2011.

F. Saleem, J. Harris, K. Zhang, and A. Harvey. Non-thermal plasma as a promising route for the removal of tar from the product gas of biomass gasification-a critical review. *Chemical Engineering Journal*, 382:122761, 2020. ISSN 1385-8947. https://doi.org/10.1016/j.cej.2019.122761.

A. Sanlisoy and M.O. Carpinlioglu. A review on plasma gasification for solid waste disposal. *International Journal of Hydrogen Energy*, 42(2):1361–1365, 2017. ISSN 0360-3199. https://doi.org/10.1016/j.ijhydene.2016.06.008.

G. Schuster, G. Loffler, K. Weigl, and H. Hofbauer. Biomass steam gasification: an extensive parametric modeling study. *Bioresource Technology*, 77(1):71–79, 2001.

N. Sharma, G. Diaz, and E. Leal-Quiros. Contact glow discharge electrolysis as an efficient means of generating steam from liquid waste. In *Proceedings of ASME IMECE2013, Paper # IMECE2013-64062*, pages 1–5, San Diego, CA, November 2013a.

N. Sharma, G. Diaz, and E. Leal-Quiros. Evaluation of contact glow-discharge electrolysis as a viable method for steam generation. *Electrochimica Acta*, 108:330–336, 2013b.

J.L. Shie, C.Y. Chang, W.K. Tu, Y.C. Yang, J.K. Liao, C.C. Tzeng, H.Y. Li, Y.J. Yu, C.H. Kuo, and L.C. Chang. Major products obtained from plasma torch pyrolysis of sunflower-oil cake. *Energy and Fuels*, 22:75–82, 2008.

N. Singh, M. Razafinimanana, and J. Hlina. Characterization of a dc plasma torch through its light and voltage fluctuations. *Journal of Physics D: Applied Physics*, 33:270–274, 2000.

J.P. Trelles, C. Chazelas, A. Vardelle, and J.V.R. Heberlein. Arc plasma torch modeling. *Journal of Thermal Spray Technology*, 18(5–6):728–752, 2009.

X. Tu, J.H. Yan, B.G. Cheron, and K.F. Cen. Fluctuations of DC atmospheric double arc argon plasma jet. *Vacuum*, 82:468–475, 2008.

G. Van Oost, M. Hrabovsky, V Kopecky, M. Konrad, M. Hlina, and T. Kavka. Pyrolysis/gasification of biomass for synthetic fuel production using a hybrid gas-water stabilized plasma torch. *Vacuum*, 83:209–212, 2009.

N. Venkatramani. Thermal plasmas in material processing. *Bulletin of Materials Science*, 18(6):741–754, 1995.

Z.A. Zainal, R. Ali, C.H. Lean, and K.N. Seetharamu. Prediction of performance of a downdraft gasifier using equilibrium modeling for different biomass materials. *Energy Conversion and Management*, 42(12):1499–1515, 2001.

5

Voltage-Enhanced Processing of Biochar*

5.1 DC Power Applied to Biochar

The main purpose is to use thermal breakdown as a mechanism for achiev-
ing conditions (temperatures) for thermochemical conversion, i.e. pyrolysis
and gasification, of carbonaceous materials. In particular for biochar, it is of
special interest to compare the magnitude of the applied voltage and power
needed to produce pyrolysis temperatures, as well as the time that such
temperatures are reached under different applied voltages. It is known that
biochar has lower electrical and thermal conductivities than graphite, but
the temperature dependence of the electrical conductivity is stronger.

5.1.1 Joule Heating of Biochar

In terms of mathematical modeling, it is important to emphasize that
because one of the main material properties used in conventional electric-
thermal models, namely the electrical conductivity, is a "macroscale"
property, it can be readily measured for most materials. The electrical
conductivity can be measured on crystalline and noncrystalline materials,
as such, this kind of models can be applied to either type of material.
In contrast, hydrodynamic models are based on "microscale" properties,
such as carrier mobilities and densities. It is very difficult to define, and
perhaps not viable to measure, these properties on noncrystalline materials
such as biochar and wood. Therefore, the versatility of an electric-thermal
model is exploited to analyze Joule heating processes in biochar.

*Corresponding Author: Williams R. Calderón-Mũnoz; Department of Mechanical
Engineering, University of Chile, Santiago, Beauchef 851, Chile.

Voltage-Enhanced Processing of Biomass and Biochar, First Edition. Gerardo Diaz.
© 2022 John Wiley & Sons Ltd. Published 2022 by John Wiley & Sons Ltd.

5.1.1.1 Case Study

A rod of carbonaceous material of length L and diameter D is subject to a constant voltage, V_{app}, applied on the left end of the rod while the right end is grounded, as shown in Figure 5.1 (Muñoz-Hernández, 2018). Due to the electric field, the motion of charge carriers produces a total current density J, which generates Joule heating in the rod that results in a temperature rise. Both sides of the rod have a fixed temperature equal to the ambient temperature T_∞. The rod is exposed to heat losses by convection to the ambient air and radiation exchange with the surrounding surfaces. The surrounding surfaces and the ambient air are at the same temperature T_∞.

A transient one-dimensional version of the electric-thermal coupled model is composed of Gauss' law (5.1a), charge conservation (5.1b), and energy conservation (5.1c), as follows:

$$\frac{\partial^2 V}{\partial x^2} = -\frac{\rho}{\epsilon_r \epsilon_0} \tag{5.1a}$$

$$\frac{\partial \rho}{\partial t} = \frac{\partial}{\partial x}\left(\sigma(T_L)\frac{\partial V}{\partial x}\right) \tag{5.1b}$$

$$C_{L,v}(T_L)\frac{\partial T_L}{\partial t} = \frac{\partial}{\partial x}\left(k_L(T_L)\frac{\partial T_L}{\partial x}\right) + \sigma(T_L)\left(\frac{\partial V}{\partial x}\right)^2 - \frac{4h_{Tot}}{D}(T_L - T_\infty) \tag{5.1c}$$

where $V(x,t)$ is the electrostatic potential, $\rho(x,t)$ is the electric charge density, and $T_L(x,t)$ is the lattice temperature. The induced magnetic field effects are small and, therefore, have been neglected. The Biot number is below 0.1 for the three materials analyzed in this chapter. Consequently, temperature gradients in the radial and angular dimension have also been neglected. Thus, the problem effectively reduces to one dimension. The

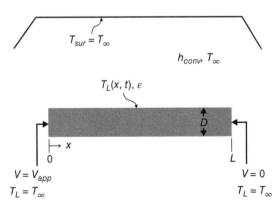

Figure 5.1 Schematic representation of the rod under an applied voltage, with boundary conditions and heat losses. Source: Muñoz-Hernández (2018) / UC Merced / Licensed under CC BY 4.0.

total heat transfer coefficient is defined as $h_{Tot} = h_{conv} + h_{rad}$, where h_{conv} is the convection coefficient, and $h_{rad} = \varepsilon \sigma_{SB}(T_{L,ave} + T_{sur})(T_{L,ave}^2 + T_{sur}^2)$ is the linearized radiation coefficient (Bergman et al., 2011). The physical parameters are the emissivity, ε, Stefan–Boltzmann's constant, σ_{SB}, the relative permittivity, ϵ_r, the permittivity of free space, ϵ_0, the electrical conductivity, $\sigma(T_L)$, the thermal conductivity, $k_L(T_L)$, and the volumetric heat capacity, $C_{L,v}(T_L) = \rho_m c_p(T_L)$, where ρ_m is the mass density and $c_p(T_L)$ is the specific heat.

The boundary conditions are $T_L(0, t) = T_L(L, t) = T_\infty$ and $V(0, t) = V_{app}$ and $V(L, t) = 0$, as seen in Figure 5.1. The initial conditions are as follows: the initial lattice temperature across the rod is equal to the ambient temperature, i.e. $T_L(x, 0) = T_{L,i} = T_\infty = 298$ K. The charge density is equal to zero everywhere, i.e. $\rho(x, 0) = 0$, which gives rise to an initial linear voltage distribution $V(x, 0) = V_{app}(1 - x/L)$ and a constant electric field $E_0 = V_{app}/L$. The initial charge density may take moderate nonzero values, generating a nonuniform initial electric field distribution; however, this moderate change in initial conditions does not affect the final results. A value of charge density equal to zero was chosen for simplicity. Note that no boundary conditions are needed for the charge density.

In this model, the current density is assumed to be uniform across the area of the rods. This assumption is more realistic for graphite than for biochar or wood due to the fibrous structure of these two materials, which affects the contact between the material and the electrode. However, it constitutes a good approximation during the onset of thermal runaway. The situation changes significantly for wood, a dielectric material, when thermal breakdown occurs, since this mechanism typically involves the formation of highly conductive channels.

The dynamics of charge and heat transport in biochar were analyzed using the following parameters: $V_{app} = 21$ V, $L = 10^{-2}$ m, and $D = 10^{-3}$ m, where the results are shown in Figure 5.2. Because of the strong temperature dependence of the electrical conductivity, the steepness of the curves near the boundaries is very large for all the variables. Similar to the case for graphite when $V_{app} = 30$ V, in this case most of the charge accumulates at the boundaries of the rods, reaching a charge density of 5.21×10^{17} m^{-3}, as shown in Figure 5.2e. The charge accumulation produces an electric field of 9180 V/m at the boundary with a minimum value of 1970 V/m inside the domain, a value slightly lower than the nominal value of 2100 V/m seen in Figure 5.2c. For the same reason, the voltage distribution also undergoes a strong distortion near the boundaries, as observed in Figure 5.2a. The simulations of biochar were performed until a maximum lattice temperature of $T_{L,max} = 100\,^{\circ}$C was obtained because of the high uncertainty in the

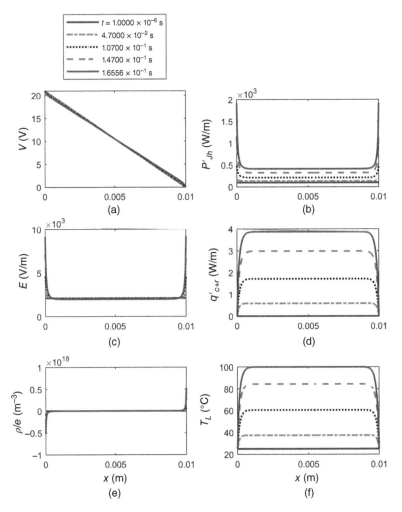

Figure 5.2 Spatial distribution and temporal evolution of various variables for biochar. Parameter values: $V_{app} = 21$ V, $L = 10^{-2}$ m, $D = 10^{-3}$ m. (a) V = electrostatic potential, (b) E = electric field, (c) P'_{jh} = Joule heating power per unit of length, (d) q'_{c+r} = heat loses due to convection and radiation per unit of length, (e) T_L = lattice temperature, (f) e = elementary charge.

properties of this material at higher temperatures. This maximum lattice temperature was reached after 1.6556×10^{-1} seconds, as seen in Fig. 5.2f. The heat losses by convection and radiation were small, reaching a maximum value of 3.9 W/m, shown in Figure 5.2d, while Joule heating power reached a much higher value of 413 W/m inside the rod and 1920 W/m at the boundaries. The value of the Joule heating power inside the domain is more than two orders of magnitude larger than the maximum heat losses by convection and radiation at the same location along the rod. In addition, the current density starts off at a value of $6.00 \times 10^4 \, \text{A/m}^2$ at room temperature and steadily increases to a maximum value of $2.66 \times 10^5 \, \text{A/m}^2$ when $T_{L,max}$ reaches $100\,°\text{C}$.

5.1.2 Joule Heating of Activated Carbon

Since we are interested in reaching very high temperatures, temperature-dependent material properties should be considered in the modeling. This is not done very commonly in the literature. As biomass is carbonized, its electrical conductivity increases by several orders of magnitude. In a similar manner, graphite is made when a mixture of hydrocarbons are heated to even higher (graphitization) temperatures. As opposed to biochar, graphite may be a highly crystalline material with a comparatively low porosity (Muñoz-Hernández, 2018).

5.1.2.1 Case Study

A drift-diffusion steady-state one-dimensional mathematical model is applied to describe the charge carrier transport and lattice temperature distribution in a graphite rod of length L and diameter D, with $L \geq D$. Voltage drop, temperature changes, and transport of electrons and holes vary only along the x-direction, as shown in Figure 5.3 (Muñoz-Hernández, 2018). A DC voltage, V_{app}, is applied on the left end of the rod while the right end is grounded. Due to the electric field, charge carriers, i.e. electrons and holes, move across the rod producing a total current density J, which

Figure 5.3
Schematic: Graphite rod under an applied voltage exposed to cooling by free convection and radiation. Source: Muñoz-Hernández (2018) / UC Merced / Licensed under CC BY 4.0.

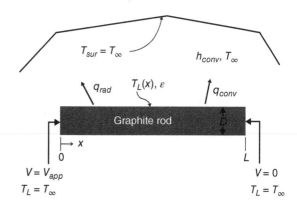

generates Joule heating in the rod. Consequently, due to Joule heating, the lattice temperature rises and results in thermal generation of electrons and holes. Both ends of the rod have a fixed temperature equal to T_∞. The rod is exposed to heat losses by natural convection to the surrounding air and radiation exchange with the surrounding surfaces; the surfaces and the air are at the same temperature T_∞.

The set of equations is composed of the following: the potential distribution described by Gauss' law in Eq. (5.2a), the continuity equations for electrons and holes, Eqs. (5.2b) and (5.2c), the momentum equations for electrons and holes in Eqs. (5.2d) and (5.2e), and the equation for the lattice temperature (5.2f)

$$\frac{d^2V(x)}{dx^2} = -\frac{e}{\epsilon_r \epsilon_0}(n_h(x) - n_e(x)) \tag{5.2a}$$

$$\frac{d\left(-en_e(x)u_e(x)\right)}{dx} = -eG_{e,net}(x) \tag{5.2b}$$

$$\frac{d\left(en_h(x)u_h(x)\right)}{dx} = eG_{h,net}(x) \tag{5.2c}$$

$$u_e(x) = \mu_e(T_L)\frac{dV(x)}{dx} - \frac{\mu_e(T_L)k_BT_c}{e}\frac{1}{n_e(x)}\frac{dn_e(x)}{dx} \tag{5.2d}$$

$$u_h(x) = -\mu_h(T_L)\frac{dV(x)}{dx} - \frac{\mu_h(T_L)k_BT_c}{e}\frac{1}{n_h(x)}\frac{dn_h(x)}{dx}$$

$$\tag{5.2e}$$

$$0 = \frac{d}{dx}\left(k_L(T_L)\frac{dT_L(x)}{dx}\right) - \frac{4h_{Tot}}{D}\left(T_L(x) - T_\infty\right)$$

$$- e\left(-n_e(x)u_e(x) + n_h(x)u_h(x)\right)\frac{dV(x)}{dx} \tag{5.2f}$$

where $V(x)$ is the electrostatic potential, $n_e(x)$ is the electron density, $n_h(x)$ is the hole density, $u_e(x)$ is the electron velocity, $u_h(x)$ is the hole velocity, $T_L(x)$ is the lattice temperature, T_∞ is the ambient temperature, and $G_{e,net}(x)$ and $G_{h,net}(x)$ are the net generation rates of electrons and holes, respectively. Lattice temperature changes only occur along the x-direction. This approximation is valid because the Biot number is lower than 0.1 (Bergman et al., 2011). It is also assumed that the electrons and holes are in thermal equilibrium (Breusing et al., 2009, Osses-Márquez and Calderón-Muñoz, 2014), for which the temperature will be referred to as the carrier temperature, T_c. It is further assumed that T_c is constant throughout the rod, but it can be set to different values. Further details regarding these assumptions are provided in Muñoz-Hernández (2018). Having a carrier temperature different from the lattice temperature allows the analysis of nonequilibrium transport. It will be shown, however, that under the assumptions and operating conditions relevant to this study, the value of T_c does not have a significant effect on the results.

The physical parameters are the elementary charge, e, the permittivity of free space, ϵ_0, the relative permittivity (dielectric constant) of graphite, ϵ_r, Boltzmann constant, k_B, the mobility of electrons, $\mu_e(T_L)$, the mobility of holes, $\mu_h(T_L)$, and the thermal conductivity of graphite, $k_L(T_L)$, where the carrier mobilities and the thermal conductivity depend on the lattice temperature. The total heat transfer coefficient is defined as $h_{Tot} = h_{conv} + h_{rad}$, where h_{conv} is the convection coefficient and $h_{rad} = \varepsilon \sigma_{SB}(T_{L,ave} + T_\infty)(T_{L,ave}^2 + T_\infty^2)$ is the linearized radiation coefficient; ε is the emissivity of graphite and σ_{SB} is the Stefan–Boltzmann constant (Bergman et al., 2011).

The current densities for electrons, $J_e(x)$, holes, $J_h(x)$, and the total current density, J, are defined as follows:

$$J_e(x) = -en_e(x)u_e(x) \tag{5.3a}$$

$$J_h(x) = en_h(x)u_h(x) \tag{5.3b}$$

$$J = J_e(x) + J_h(x) \tag{5.3c}$$

Global current continuity requires the total current density to be constant, thus the following relationship holds

$$\frac{dJ}{dx} = \frac{d(J_e(x) + J_h(x))}{dx} = 0 \tag{5.4}$$

In addition, the electric field is related to the potential as

$$E(x) = -\frac{dV(x)}{dx} \tag{5.5}$$

The net generation rate of electrons is equal to the net generation rate of holes (Volovichev et al., 2009), that is

$$G_{e,net}(x) = G_{h,net}(x) = G_{net}(x) \tag{5.6}$$

This means that when an electron gains sufficient energy to jump from the valence band to the conduction band, a hole is created and left behind in the valence band.

Because we do not know the net generation rates explicitly, we take a different approach in solving the continuity equations for the electron and hole densities. Substituting Eqs. (5.2d) and (5.2e) into Eqs. (5.3), and rearranging, we get

$$J = -e\left(n_e(x)\mu_e(T_L) + n_h(x)\mu_h(T_L)\right)\frac{dV(x)}{dx}$$
$$+ k_B T_c\left(\mu_e(T_L)\frac{dn_e(x)}{dx} - \mu_h(T_L)\frac{dn_h(x)}{dx}\right) \tag{5.7}$$

Since intrinsic graphite is being used, the electron and hole mobilities are replaced with an average carrier mobility, $\mu_{ave}(T_L)$, hereafter called mobility. Thus,

$$J = 2e\mu_{ave}(T_L)n_{th}(T_L)E(x) - k_B T_c\mu_{ave}(T_L)\frac{d}{dx}\left(n_h(x) - n_e(x)\right) \tag{5.8}$$

where we have used the expression $2\mu_{ave}(T_L)n_{th}(T_L)$ in place of $n_e(x)\mu_e(T_L) + n_h(x)\mu_h(T_L)$; $n_{th}(T_L)$ is the thermal or intrinsic density. Thermal density and intrinsic density refer to the same parameter, and will be used interchangeably in this dissertation.

Graphite is sometimes referred to as a zero-gap semiconductor (Wallace, 1947) or as a semimetal (Klein, 1962) because it does not have a band gap. Instead graphite has a small overlap between the valence band and the conduction band (McClure, 1957, Klein, 1962, 1964). The presence of a band overlap indicates that the electron and hole densities are nearly equal to each other (and to the thermal density) (Klein, 1962, 1964). Based on the two-band model, the thermal density can be approximated as (Klein, 1964)

$$n_{th}(T_L) = \frac{16\pi m_c^{\star}}{h_P^2 c_0}k_B T_L \ln\left[1 + \exp\left(\frac{\mathcal{E}_{bo}}{2k_B T_L}\right)\right] \tag{5.9}$$

where $m_c^{\star}$ is the effective carrier mass, c_0 is twice the layer spacing, h_P is Planck's constant, and $\mathcal{E}_{bo}$ is the band overlap energy.

The first term on the right-hand side of Eq. (5.8) is the drift current density, J_{Drift}, and the second term is the diffusion current density, J_{Diff}. Under the conditions used here, and using graphite properties, the drift current density becomes much larger than the diffusion current density for all physically meaningful values of the carrier temperature T_c. Thus, the diffusion term can be safely neglected, and the total current density is fully determined by the drift current density.

$$J \approx 2e\mu_{ave}(T_L)n_{th}(T_L)E(x) \tag{5.10}$$

The group of parameters multiplying the electric field in Eq. (5.10) is defined as the electrical conductivity, $\sigma = 2e\mu_{ave}n_{th}$, and the well-known expression for current density may be obtained, $J = \sigma E$.

Making use of subsequent expressions, Eqs. (5.2) can be simplified and recast as follows:

$$J = 2e\mu_{ave}(T_L)n_{th}(T_L)E(x) \tag{5.11a}$$

$$0 = \frac{d}{dx}\left(k_L(T_L)\frac{dT_L(x)}{dx}\right) - \frac{4h_{Tot}}{D}\left(T_L(x) - T_\infty\right) + JE(x) \tag{5.11b}$$

$$E(x) = -\frac{dV(x)}{dx} \tag{5.11c}$$

$$\frac{d^2V(x)}{dx^2} = -\frac{e}{\epsilon_r\epsilon_0}(n_h(x) - n_e(x)) \tag{5.11d}$$

The boundary conditions are as follows: $V = V_{app}$ at $x = 0$, $V = 0$ at $x = L$, and $T_L = T_\infty$ at $x = 0, L$, as shown in Figure 5.3.

Experimental data suggests that the electron, hole, and thermal densities are nearly equal to each other (Klein, 1964), which may lead us to assume that they are exactly equal. It will be shown that for the biased intrinsic graphite rods, all three carrier densities are nearly equal to each other, but not exactly equal. Moreover, assuming their equality does not lead to a consistent solution. In addition, we will show that the maximum absolute difference among the carrier densities is several orders of magnitude smaller than the thermal density.

In the absence of a gate voltage, the small differences among the electron, hole, and thermal densities in intrinsic graphite are important because, though small, they give rise to non-constant electric fields. On the contrary, in graphene transistors, for example, the gate voltage gives rise to large differences among the electron, hole, and thermal densities, which results in very large variations in the electric field (Bae et al., 2010, Freitag et al., 2010, Bae et al., 2011). Therefore, the small differences in carrier densities of the intrinsic material become insignificant.

The properties of pyrolytic graphite are presented in Table 5.1. The experimental data for the temperature-dependent mobility (Klein, 1964), from room temperature to 1000 K, was fitted with the following function, with an R^2 value of 0.9954, in order to use it in the simulations

$$\mu_{ave}(T_L) = 7211.2T_L^{-1.555} \tag{5.12}$$

Similarly, the experimental data for the temperature-dependent thermal conductivity (Ho et al., 1972, Fugallo et al., 2014), was fitted with the following function, with an R^2 value of 0.9988,

$$k_L(T_L) = 839269.0T_L^{-1.068} \tag{5.13}$$

Table 5.1 Properties of pyrolytic graphite.

Property	Symbol	Value
Twice the graphene layer spacing	c_0	0.672×10^{-9} m
Band overlap energy	$\mathcal{E}_{bo}$	0.01 eV
Emissivity	ε	0.8
Relative permittivity	ϵ_r	13.0
Free electron mass	m_0	9.109×10^{-31} kg
Effective carrier mass	$m_c^\star/m_0$	0.0125
Thermal conductivity at $T_L = 298$ K	$k_{L,ref}$	1911 W/(m K)
Average carrier mobility at $T_L = 298$ K	$\mu_{ave,ref}$	1.02 m^2/(V s)

Sources: Klein (1964), Ho et al. (1972), Fugallo et al. (2014), and Entegris (2015).

Table 5.2 System parameter values.

Parameter	Symbol	Value
Rod length	L	10^{-2} m
Rod diameter	D	10^{-3} m
Convection heat transfer coefficient	h_{conv}	10 W / (m^2 K)
Ambient temperature	T_∞	298 K

A value for the dielectric constant of graphite for a constant (DC) voltage was not found, but values were found for exfoliated graphite for frequencies as low as 50 Hz (Hong and Chung, 2015). The values for washed and unwashed exfoliated graphite samples at 50 Hz were 38 and 364, respectively (Hong and Chung, 2015). Nonetheless, there is an uncertainty as to how the dielectric constant changes at lower frequencies and for a DC voltage. On the other hand, dielectric constant values for conventional semiconductors are well known. For example, the values for Si and GaAs are 11.9 and 12.9, respectively (Sze and Ng, 2006).

Based on this information, a dielectric constant of 13 will be used for graphite throughout this chapter.

Unless otherwise stated, the system parameter values used for cases *I.A,B*, and for some of the other cases, are listed in Table 5.2.

In the published literature, several approximations for the intrinsic density of pyrolytic graphite have been proposed based on available experimental data. A popular expression to calculate the intrinsic density of pyrolytic graphite, Eq. (5.9), has been developed by Klein (1964) according to his two-band theory. The fact that there is a band overlap rather than a band gap means that there are nearly equal numbers of electron and holes. As a result, it has been generally assumed that the electron and hole densities are exactly equal. In this section, several cases are solved to demonstrate that although the electron and hole densities are nearly equal to each other and to the thermal density, there exists a small difference among them. In addition, the difference is also bounded to a maximum value dictated by material properties, geometry, and operating conditions. The order of magnitude of the maximum absolute difference between holes and electrons is here quantified with the aid of Gauss' law in the form of the Poisson equation.

The consistency in the solution of the governing equations is analyzed based on the assumption of equality of the electron, hole, and thermal densities. From Eq. (5.4), we deduce that the total current density, J, is constant. Also, in Eqs. (5.11a) and (5.11b), the thermal density $n_{th}(T_L)$ and carrier mobility $\mu_{ave}(T_L)$ are dependent on the lattice temperature $T_L(x)$, and the electric field $E(x)$ is a function of position. In order to solve the system of equations, Eqs. (5.11) were discretized using a second order finite difference numerical scheme with a mesh of 101 nodes. Convergence was achieved when the maximum absolute relative error for temperature falls below a tolerance of $\delta = 10^{-5}$, i.e. $\max|T_{L,new} - T_{L,old}|/T_{L,new} \leq \delta$. The appropriate grid size and tolerance value were chosen based on a grid independence analysis.

Case I.A: $D = 10^{-3}$ m, $L = 10^{-2}$ m, and $V_{app} = 1.0$ V; assume $n_e(x) = n_h(x) \approx n_{th}(T_L)$. When $n_e(x) = n_h(x)$, the right side of Poisson's equation, Eq. (5.11d), becomes equal to zero, and Poisson's equation reduces to Laplace's equation, $d^2V/dx^2 = 0$. Solution of the Laplace equation leads to a linear voltage distribution, and consequently to a constant electric field $E(x) = E_0 = V_{app}/L$. The thermal density can be obtained from two expressions, i.e. directly from Eq. (5.9), or by rewriting Eq. (5.11a) in the form:

$$\tilde{n}_{th}(x) = \frac{J}{2e\mu_{ave}(T_L)E_0} \tag{5.14}$$

where it has been labeled as $\tilde{n}_{th}$ to differentiate from the value of n_{th} obtained from Eq. (5.9).

The results obtained for this case are shown in Figure 5.4, where the current density was obtained iteratively by evaluating the thermal density and the mobility at the average lattice temperature, and using a constant electric field, i.e. $J = 2e\,\mu_{ave}(T_{L,ave})n_{th}(T_{L,ave})E_0$. As expected, Figure 5.4a shows a constant electric field and a linear variation of the voltage due

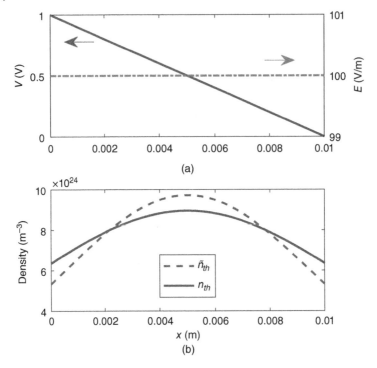

(a)

(b)

Figure 5.4 Results for *Case I.A*: $n_e(x) = n_h(x)$. Parameter values: $D = 10^{-3}$ m, $L = 10^{-2}$ m, $V_{app} = 1.0$ V.

to the assumption of $n_e(x) = n_h(x)$. However, Figure 5.4b shows that the values of n_{th} and $\tilde{n}_{th}$ are not equal, except in a couple of locations where the curves intersect each other, and thus, under this assumption, the system of equations is inconsistent. This solution suggests that the electric field should not be constant; and therefore, the electron, hole, and thermal densities should be approximately equal but not exactly equal. Under these assumptions, the only possibility for the electric field to be constant would be if $n_{th}(T_L)$ and $\mu_{ave}(T_L)$ had an exactly inversely proportional dependence on the lattice temperature $T_L(x)$. However, this is not the case. The thermal density increases with temperature in a nearly linear fashion, while the mobility decreases much faster with temperature, following a power law.

The results shown suggest that while the hole and electron densities are nearly equal to each other and to the thermal density, i.e. $n_e(x) \approx n_h(x) \approx n_{th}(T_L)$, there is a small difference between the hole and electron densities $|n_h(x) - n_e(x)| \ll n_{th}(T_L)$. One of the main objectives in this chapter is to numerically quantify the difference between hole and electron densities $n_h(x) - n_e(x)$, which may change along the rod. The first step is to quantify the maximum absolute difference, $\max|n_h(x) - n_e(x)|$, which may occur anywhere along the rod.

To simplify the analysis, the electric potential and the position parameter x will be nondimensionalized as follows: $V^* = V/V_{app}$, and $x^* = x/L$. Thus, Poisson's equation (5.11d) becomes

$$\frac{d^2V^*(x^*)}{dx^{*2}} = -\frac{eL^2}{\epsilon_r\epsilon_0 V_{app}} \max|n_h(x^*) - n_e(x^*)| \tag{5.15}$$

In order to quantify the maximum difference between hole and electron densities, only the magnitude of this value is considered, which is independent of position. Integrating Eq. (5.15) analytically from 0 to 1 yields the equation for a parabola

$$V^*(x^*) = \frac{eL}{2\epsilon_r\epsilon_0 E_0} \max|n_h(x^*) - n_e(x^*)| \left[x^* - x^{*2}\right] + \left[1 - x^*\right] \tag{5.16}$$

where $E_0 = V_{app}/L$.

Typical values for this problem are taken as length $L = 10^{-2}$ m and electric field of 100 V/m, where the dielectric constant used for graphite is 13.0. Substituting and simplifying, the following expression is obtained:

$$V^*(x^*) = (6.96 \times 10^{-14} \text{m}^3) \max|n_h(x^*) - n_e(x^*)| \left[x^* - x^{*2}\right] + \left[1 - x^*\right] \tag{5.17}$$

Because the units of the electron/hole densities are in m^{-3}, the units will cancel out upon multiplication. Although integration of the Poisson equation yields a parabolic equation, a nearly linear voltage distribution is expected. This means that the coefficient multiplying the quadratic term should be of order 1, nearly equal to the linear term. In order for this coefficient to have such magnitude, the maximum absolute difference between holes and electrons, $\max|n_h(x) - n_e(x)|$, must be of the order of 10^{14} m^{-3} or lower. The potential distribution using Eq. (5.17) was plotted for various values of $\max|n_h(x) - n_e(x)|$ in Figure 5.5a. It can be seen that as $\max|n_h(x) - n_e(x)|$ increases, the voltage distribution becomes more parabolic than linear, and the peak voltage is higher than the applied voltage. When the peak voltage becomes higher than the applied voltage, it is considered as a nonphysical solution. In order to obtain a meaningful physical solution for the given length $L = 10^{-2}$ m and applied voltage $V_{app} = 1.0$ V, the maximum absolute difference, $\max|n_h(x) - n_e(x)|$, should be in the order of 10^{13} m^{-3}. As expected, this number is much smaller than the thermal density. Based on this analysis, it can be estimated that in order

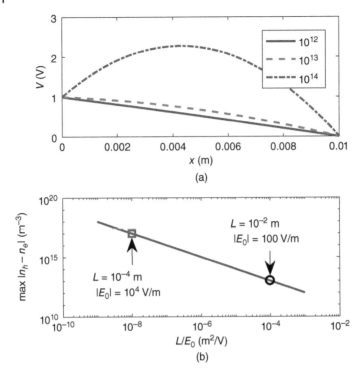

Figure 5.5 (a) Voltage distribution for various values of $\max|n_h(x) - n_e(x)|$ obtained with Eq. (5.17). Parameter values: $D = 10^{-3}$ m, $L = 10^{-2}$ m, $V_{app} = 1.0$ V, ($E_0 = 100$ V/m); legend: $\max|n_h(x) - n_e(x)|$ (m^{-3}). (b) $\max|n_h(x) - n_e(x)|$ vs. L/E_0.

to obtain a meaningful physical solution for the voltage distribution, the maximum difference between holes and electrons is limited to a certain value. This value is dictated directly by the dielectric constant of the material ϵ_r, the applied voltage V_{app}, the length L, and indirectly affected by the carrier mobility μ_{ave}, the effective carrier mass $m_c^\star$, and diameter D.

On the other hand, in order to have a slightly nonlinear voltage distribution, there needs to be a small difference between the electron and hole densities. Moreover, note that the electron, hole, and thermal densities vary along the rod due to a nonuniform temperature distribution; therefore, it is expected that the difference $n_h(x) - n_e(x)$ also varies along the length of the rod. In conclusion, it can be stated that for a given ϵ_r, knowing that $n_h(x) \approx n_e(x) \approx n_{th}(T_L)$, it is expected that $0 < \max|n_h(x) - n_e(x)| < \xi(L/E_0)$, where $\xi(L/E_0)$ is a function of L/E_0.

Poisson's equation can also be written incorporating the Reynolds number, Re, which is analogous to the Reynolds number used in fluid mechanics.

Rewriting Eq. (5.15), we obtain

$$\frac{d^2 V^*(x^*)}{dx^{*2}} = -\frac{\mu_{ave}^2 m_c^\star}{\epsilon_r \epsilon_0} \frac{1}{Re} \max|n_h(x^*) - n_e(x^*)| \qquad (5.18)$$

where the Reynolds number is defined as (Osses-Márquez and Calderón-Muñoz, 2014)

$$Re = \frac{\mu_{ave}^2 m_c^\star (|V_{app}|/L)}{eL} \qquad (5.19)$$

In fluid mechanics, the Reynolds number quantifies the ratio of the inertia to the viscous forces; as Re increases the inertia forces dominate. Based on Eq. (5.19), for the given material properties μ_{ave}, m_{eff}, and ϵ_r, Re increases with E_0/L. Thus, as a consequence, $\max|n_h(x) - n_e(x)|$ increases with Re.

Before wrapping up this section, it is noted that the results were obtained using a value of 13 for the dielectric constant, ϵ_r. Because this value was not found, let us explore what happens when this value changes in magnitude. The dielectric constant is in the denominator on the coefficient multiplying the first term on the right-hand side of Eq. (5.16). Therefore, as ϵ_r increases, the coefficient decreases and $\max|n_h(x) - n_e(x)|$ increases. For instance, if $\epsilon_r = 1300$ instead of 13, the coefficient in Eq. (5.17) is decreased by two orders of magnitude to 6.96×10^{-16} m^3. As a consequence, $\max|n_h(x) - n_e(x)|$ increases by two orders of magnitude as well.

The previous analyses showed that indeed the electron, hole, and thermal densities are nearly equal, yet slightly different. The order of magnitude of the maximum absolute difference between hole and electron densities was quantified under different operating conditions. The governing equations given by Eqs. (5.11) are solved consistently, i.e. the electric field is not constant due to the fact that the electron and hole densities are not exactly equal. The system of equations (5.11) is rewritten for convenience, and rearranged to resemble the numerical algorithm as follows:

$$E(x) = \frac{J}{2e\mu_{ave}(T_L)n_{th}(T_L)} \qquad (5.20a)$$

$$0 = \frac{d}{dx}\left(k_L(T_L)\frac{dT_L(x)}{dx}\right) - \frac{4h_{Tot}}{D}[T_L(x) - T_\infty] + JE(x) \qquad (5.20b)$$

$$-\frac{dV(x)}{dx} = E(x) \qquad (5.20b)$$

$$n_h(x) - n_e(x) = \frac{\epsilon_r \epsilon_0}{e}\frac{dE(x)}{dx} = -\frac{\epsilon_r \epsilon_0}{e}\frac{d^2 V(x)}{dx^2} \qquad (5.20c)$$

The equations were discretized using finite differences. The algorithm is as follows: given an initial value for the total current density, Eqs. (5.20a) and (5.20b) are solved iteratively until convergence for temperature is obtained. Once the temperature converges, Eq. (5.20c) is integrated to determine the voltage distribution, and the voltage solution at the left boundary is compared to the applied voltage. If the solution voltage at the left end differs from the applied voltage by more than the given tolerance, $|(V_{app} - V(0))/V_{app}| \leq \delta$, the current density is increased/decreased accordingly; this procedure is a form of the shooting method. Equations (5.20a) through (5.20b) are then solved iteratively until convergence for both temperature and voltage is achieved. Consequently, Eq. (5.20d) is used to solve for $n_h(x) - n_e(x)$.

Case I.B: $D = 10^{-3}$ m, $L = 10^{-2}$ m, and $V_{app} = 1.0$ V. Results for this case are shown in Figure 5.6, where it can be seen that the electric field is not constant, and its average magnitude is close to the nominal value $E_0 = V_{app}/L = 100$ throughout the rod. The fact that the electric field is not constant gives rise to a nonlinearity in the voltage distribution, which can be seen as a slight distortion on the almost linear solid line, shown in Figure 5.6a. The

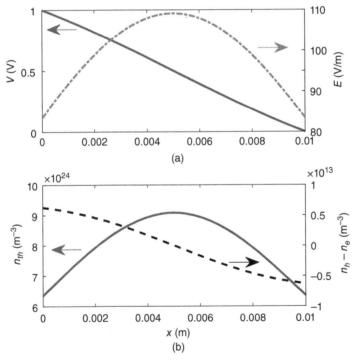

Figure 5.6 Results for *Case I.B* with parameter values: $D = 10^{-3}$ m, $L = 10^{-2}$ m, $V_{app} = 1.0$ V.

thermal density and the difference between the hole and electron densities are shown in Figure 5.6b. Because the thermal density has a nearly linear dependence with respect to temperature, the thermal density curve resembles that of the temperature distribution, reaching a maximum value slightly higher than 9×10^{24} m^{-3}. The difference between hole and electron densities is represented by the dashed black curve. It can be seen that the values vary approximately between $+0.5 \times 10^{13}$ m^{-3} and -0.5×10^{13} m^{-3}, displaying a non-symmetric distribution, where the difference becomes zero at the midpoint along the length of the rod. It is also observed that the maximum difference between holes and electrons coincides with the largest gradient of electric field and temperature, which in this case occurs near the boundaries. In addition, there are more holes than electrons on the left half of the rod due to the positive voltage and more electrons than holes on the right half of the rod near the grounded end. Although $\max|n_h(x) - n_e(x)| \ll n_{th}(T_L)$, the small difference gives rise to the small non-linearity in the voltage and a non-constant electric field. The operating conditions used for this case corresponds to the black circle depicted in Figure 5.5b.

In addition to the results previously shown, Eq. (5.20b) is used to quantify the Joule heating power and the heat losses. Moreover, Eqs. (5.2b) and (5.2c) are used to compute the net generation terms for electrons and holes, respectively, and the second term on the right-hand side of Eq. (5.8) is used to compute the diffusion current density, J_{Diff}.

In this case, the parameter values from Table 5.2 are used together with an applied voltage, V_{app}, equal to 1.0 V to obtain the distribution of the lattice temperature, electric field, electron and hole densities, as well as, the Joule heating effect. These parameter values correspond to those used in *Case I.A*. For consistency across the figures in this section, the results for *Case I.B* shown in Figure 5.6 are included in Figure 5.7. Figure 5.7c shows the Joule heating power and heat losses due to convection and radiation, given in a per-unit-length basis.

The Joule heating power, P'_{Jh}, has an order of magnitude of about 1.5×10^4 W/m, about three to four orders of magnitude higher than the heat losses, $q'_{r+c} = (4h_{Tot}(T_L(x) - T_\infty)/D)A_c$. The heat losses are small due to the relatively small surface area of the rods. The temperature distribution profile is shown in Figure 5.7d. The maximum temperature reaches about 173 °C in the middle of the rod, due to the symmetric boundary conditions. The electron and hole velocities are equal in magnitude, but have opposite signs. Their magnitude ranges from about 85 m/s at the boundaries to a minimum of about 60 m/s in the middle of the rod, see Figure 5.7e. The velocity is proportional to the mobility, and because the mobility decreases with temperature, so does the velocity. Since the mobility has

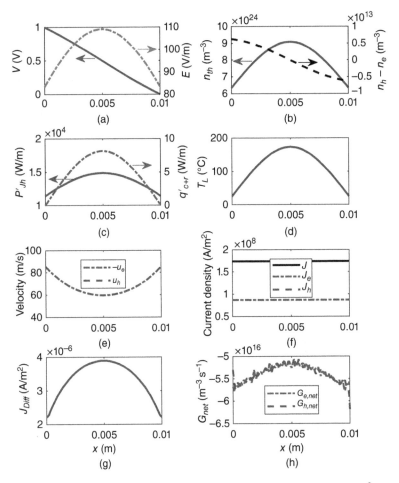

Figure 5.7 Results for *Case I.B* with parameter values: $V_{app} = 1.0\,\text{V}, L = 10^{-2}\,\text{m}$, $D = 10^{-3}\,\text{m}$.

an order of magnitude close to one, the magnitude of the velocity follows closely the magnitude of the electric field. The total, electron and hole current densities are shown in Figure 5.7f. The total current density has a constant value of about $1.74 \times 10^8\,\text{A/m}^2$. It is the sum of the electron and hole densities, which have equal values. The diffusion current density is shown in Figure 5.7g. Its value was calculated using a carrier temperature, $T_c = 298\,\text{K}$. It can be seen that its value is positive, varies with position, and is several orders of magnitude lower than the total current density. For this case, T_c would have to be several orders of magnitude higher in order for the diffusion current density to have a significant contribution to the total current density. Thus, this result corroborates the assumptions $J \approx J_{Drift}$, and $J_{Diff} \approx 0$.

The net generation rates for electrons and holes are shown in Figure 5.7h. Both values are negative and equal to each other. The value of the net generation rate is negative, meaning that recombination rate is larger than the generation rate. The absolute value of the net generation rate is in the order of 10^{16} m^{-3} s^{-1}. The curve shows a significant amount of noise because G_{net} is obtained by dividing over the elementary charge, which has an order of magnitude of 10^{-19}. Also, G_{net} is a secondary variable, i.e. it is calculated after the overall solution of the problem is obtained.

At this time, it is noted that previous results were obtained using a dielectric constant value of 13. Additional cases were run using higher values of ϵ_r to determine its effect on the overall results. However, except for increases in the difference between electrons and holes, there are no visible changes to the rest of the results.

Case II: The parameter values from Table 5.2 are used again, but with an applied voltage, $V_{app} = -2.0$ V, in order to analyze the effects of polarity and magnitude of the applied voltage. The voltage distribution and the electric field are shown in Figure 5.8a. The electric field has a larger variation, and therefore, the nonlinearity of the voltage distribution is more pronounced compared to *Case I.B*.

The thermal density and the difference between the hole and electron densities are shown in Figure 5.8b. It can be seen that the magnitude of the difference increased compared to *Case I.B*. A maximum absolute difference between holes and electrons of about 0.5×10^{13} m^{-3} was calculated for *Case I.B* compared to about 2.7×10^{13} m^{-3} for *Case II*. Also, because the polarity of the voltage changed, the variation of the difference between holes and electrons changed, there are more electrons on the left half of the rod and more holes on the right (grounded) half.

The Joule heating power and heat losses are shown in Figure 5.8c, where it is seen that the magnitude of both the Joule heating power and the heat losses increased with voltage. The maximum Joule heating power increased from 1.49×10^4 W/m for *Case I.B* to 4.86×10^4 W/m for *Case II*, while the maximum heat losses increased from 8.16 W/m for *Case I.B* to 85.4 W/m for *Case II*. Although the heat losses increased more with voltage, the order of magnitude of the Joule heating power is still about two to three orders of magnitude higher.

The temperature distribution is shown in Figure 5.8d, where the maximum temperature increased from 173 °C for *Case I.B* to 763 °C for *Case II*. The total current density for *Case II* is -2.42×10^8 A/m^2, which is about a

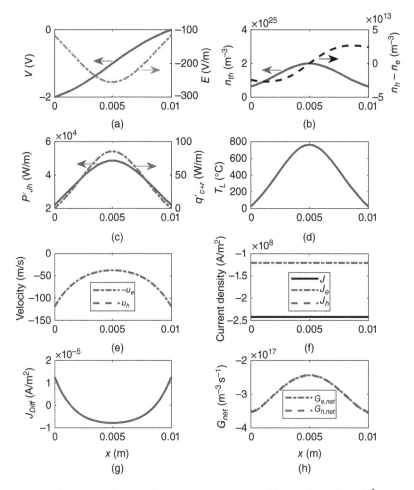

Figure 5.8 Results for *Case II* with parameter values: $V_{app} = -2.0\,\text{V}$, $L = 10^{-2}\,\text{m}$, $D = 10^{-3}\,\text{m}$.

39% increase from the *Case I.B*. This result suggests that the Joule heating power (or total current density) does not vary linearly with applied voltage.

The electron/hole velocities are shown in Figure 5.8e. Because the applied voltage changed sign from positive to negative, their values also changed sign. The magnitude of the velocities increased due to the increase in applied voltage. The total current density also changed sign due to the negative applied voltage, it is shown in Figure 5.8f. Its magnitude increased to $2.42 \times 10^{8}\,\text{A/m}^{2}$. Again, its value is the sum of the electron and hole current densities, which are equal in magnitude.

The diffusion current density is shown in Figure 5.8g. Its value varies between about $+1.26 \times 10^{-5}$ A/m^2 near the boundaries to about -7.83×10^{-6} A/m^2 at the middle. In this case, the diffusion current density became negative at and around the middle section. In addition, the order of magnitude is still similar to *Case I.B*. The net generation rate is shown in Figure 5.8h. The values still remain negative, but about one order of magnitude higher than *Case I.B*.

As discussed previously, both the thermal conductivity and mobility are dependent on the lattice temperature, and results were obtained and analyzed using the temperature-dependent values, *Cases I.B* and *II*. Several more cases are solved assuming quasi-constant values for k_L, μ_{ave}, and n_{th}. Thus, a single value of k_L, μ_{ave}, and n_{th} evaluated at the lattice average temperature, $T_{L,ave}$, will be used across the rod. To a lesser extent, this methodology still allows to take into account the effect of the increasing temperature without having to evaluate k_L and μ_{ave} at each node. Thus, saving some computational effort, especially in transient modeling. However, making such simplifications may reduce the accuracy of the results, as will be shown next.

Case I.B.1: Uses the same parameter values as in *Case I.B*, but with $k_L = k_L(T_{L,ave})$. The results for this case are shown in Figure 5.9. Comparing these results to the results from *Case I.B*, it is observed that $E(x)$, $V(x)$, $n_{th}(x)$, $P'_{Jh}(x)$, $q'_{r+c}(x)$, $u_h(x)$, and $u_e(x)$ remain nearly equal. The difference $n_h(x) - n_e(x)$ has less curvature, and its magnitude reaches slightly larger values near the boundaries, Figure 5.9b; the temperature curve is smoother, Figure 5.9d, and $T_{L,max} = 185$ °C, which is 12 degrees higher than *Case I.B*; J had a small decrease, seen in Figure 5.9f, from 1.74×10^8 A/m^2 to 1.70×10^8 A/m^2; the curvature of J_{Diff} flipped, Figure 5.9g, and its magnitude decreased by about 10 times; lastly, the value of G_{net}, shown in Figure 5.9h, increased by nearly one order of magnitude. In this case, $T_{L,max}$ reached a higher value because the thermal conductivity near the boundaries has a smaller magnitude, since k_L decreases significantly with temperature, therefore, less heat is dissipated to the boundaries. In addition, because the electrical conductivity of pyrolytic graphite decreases with temperature, the total current density also decreased by a small amount.

Case I.B.2: Uses the same parameter values as in *Case I.B*, but with $k_L = k_L(T_{L,ave})$ & $\mu_{ave} = \mu_{ave}(T_{L,ave})$. The results for this case are shown in Figure 5.10. Comparing to *Case I.B*, it is readily observed that the curvature of the electric field is flipped, Figure 5.10a, and its highest magnitude has increased from about 109 to 127 V/m. The fact that the electric field is

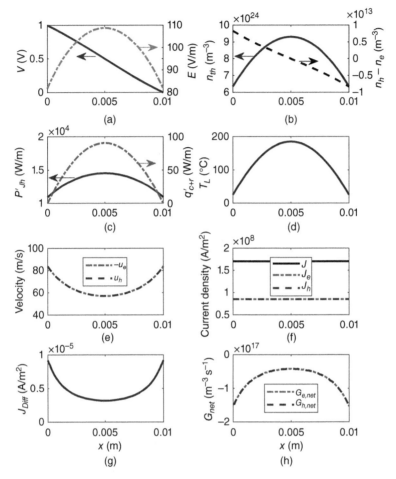

Figure 5.9 Results for *Case I.B.1*: $k_L = k_L(T_{L,ave})$, with parameter values: $V_{app} = 1.0$ V, $L = 10^{-2}$ m, $D = 10^{-3}$ m.

lower near the middle of the rod is due to the higher temperature at this location. Recall that $\sigma = 2e\mu_{ave}n_{th}$, and in this case, we have only evaluated μ_{ave} at $T_{L,ave}$, while n_{th} has been evaluated at the local temperature at each node. This means that while μ_{ave} has a constant value across the rod, the magnitude of n_{th} will increase from the boundaries to the middle of the rod. Thus, σ increases in the same manner, i.e. σ increases with temperature, which contradicts the real dependency with temperature. Since $J = \sigma E$ is constant, higher values of σ mean lower values of E. The voltage curve, Figure 5.10a has changed from a "Z" shape to an "S" shape,

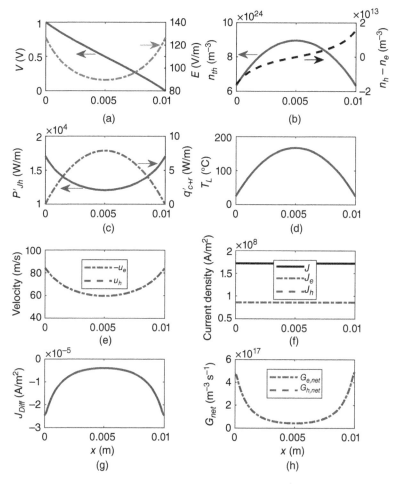

Figure 5.10 Results for *Case I.B.2*: $k_L = k_L(T_{L,ave})$, $\mu_{ave} = \mu_{ave}(T_{L,ave})$, with parameter values: $V_{app} = 1.0\,$V, $L = 10^{-2}\,$m, $D = 10^{-3}\,$m.

and the curvature is more pronounced. While the thermal density remains nearly the same, the difference between holes and electrons increases monotonically from left to right, Figure 5.10b, and the magnitude also increases by about a factor of three compared to *Case I.B*. The curvature of the Joule heating curve also flips, following the shape of the electric field, while the magnitude increases near the boundaries and decreases around the middle of the rod, Figure 5.10c; the curvature and magnitude of the heat losses remain about the same. The temperature distribution again is smoother, Figure 5.10d, and $T_{L,max}$ reaches 167 °C, which is 6 degrees lower than *Case I.B*. This occurs because the Joule heating (and electric field) has a smaller magnitude than *Case I.B* near the middle of the rod. The carrier velocities, $u_h(x)$ & $u_e(x)$, Figure 5.10e, are also nearly equal to *Case I.B*. The velocities depend on both the mobility and electric field, and because the mobility is constant, the velocities have a scaled curvature of the electric field. The total current density, $J = 1.72 \times 10^8$ A/m^2, Figure 5.10f, which is slightly smaller than *Case I.B*. The diffusion current density has changed sign from positive to negative, Figure 5.10g, and the magnitude is one order smaller than *Case I.B*. Lastly, the net generation rate is positive, Figure 5.10h, with an order of magnitude of 10^{17} m^{-3} s^{-1}.

Case I.B.3: Uses the same parameter values as in *Case I.B*, but with $k_L = k_L(T_{L,ave})$, $\mu_{ave} = \mu_{ave}(T_{L,ave})$, and $n_{th} = n_{th}(T_{L,ave})$. The fact that both the mobility and the thermal density have a constant value across the rod means that the electrical conductivity also has a constant value across the rod. This causes several curves to also be constant across the rod, such is the case for $E(x)$, $n_h(x) - n_e(x) = 0$, $P'_{Jh}(x)$, $u_h(x)$ & $u_e(x)$, $J_{Diff} = 0$, and $G_{net} = 0$, as shown in Figure 5.11. A constant electric field yields a linear voltage distribution as seen on Figure 5.11a. A constant electric field also creates a difference between electrons and holes equal to zero, Figure 5.11b, though the curve shows a small value near the boundaries due to numerical error. The Joule heating power has constant value of 1.34×10^4 W/m, Figure 5.11c, which appears to be the mean power when compared to *CaseI.B*. The heat losses, Figure 5.11c, remain about equal because the lattice temperature is nearly the same as *Case I.B*. The lattice temperature, Figure 5.11d, reaches a maximum value of 178 °C, 5 degrees hotter than *Case I.B*. The carrier velocities, Figure 5.11e, have a constant value of 65.1 m/s, which is 5 m/s faster than the minimum velocity from *Case I.B*. The current density, Figure 5.11f, has a value of $J = 1.71 \times 10^8$ A/m^2, compared to $J = 1.74 \times 10^8$ A/m^2 from *Case I.B*. The diffusion current density and the net generation rate are both zero, Figure 5.11g,h, respectively.

While using a constant value of the thermal conductivity only shifts the curves slightly, evaluating the mobility, or both the mobility and the thermal density, at the average lattice temperature affects the results quite significantly. However, the temperature distribution and the total current density remain relatively equal to those values from *CaseI.B*. If these are the only results of interest, then the quasi-constant values for the mobility and thermal conductivity can be used to produce the desired results knowing that the results will contain some error. However, if one desires an accurate account of the rest of the parameters, one needs to evaluate the temperature-dependent values at each node.

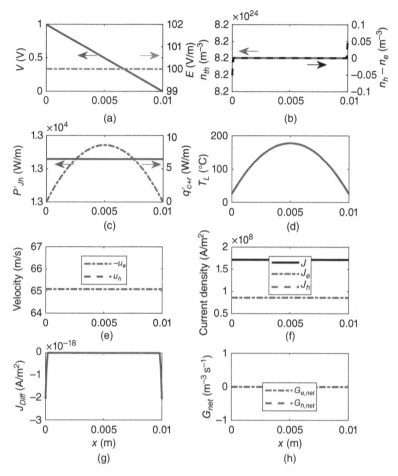

Figure 5.11 Results for *Case I.B.3*: $k_L = k_L(T_{L,ave})$, $\mu_{ave} = \mu_{ave}(T_{L,ave})$, $n_{th} = n_{th}(T_{L,ave})$ with parameter values: $V_{app} = 1.0$ V, $L = 10^{-2}$ m, $D = 10^{-3}$ m.

5.1.3 Recent Trends in Mathematical Modelling

Semiconductor devices such as transistors are based in the manipulation of charge carrier (electron/hole) transport. Such semiconductor devices usually consist of a single charge carrier, typically electrons. As a consequence, the majority of the applications of hydrodynamic models are dedicated to analyze the flow of electrons (Bløtekjær, 1970, Majumdar et al., 1995b, Grondin et al., 1999, Mohseni et al., 2005, Calderón-Muñoz, 2009). Nonetheless, the flow of holes is also important in silicon and other semiconductor devices (Jonscher, 1964, Cook, 1983, Wang, 1985,

Aluru et al., 1993, Lai and Majumdar, 1996, Smith and Brennan, 1998, Ballestra et al., 2002). In modern times, in addition to conventional semiconductor devices, upon its discovery (Novoselov et al., 2004), graphene has caught the attention of many researchers because of its potential applications in electronic devices (Meric et al., 2008, Bae et al., 2010, Freitag et al., 2010, Dorgan et al., 2010, Arora et al., 2012, Neto et al., 2009). In graphene devices, the flow of holes may be as important as the flow of electrons.

The main reasons to use hydrodynamic models are two: (i) to numerically quantify the difference between the electron, hole, and thermal densities; and (ii) to analyze charge transport and Joule heating in intrinsic pyrolytic graphite rods.

5.1.3.1 Transient One-Dimensional Hydrodynamic Model

The one-dimensional (x-direction) transport of electrons and holes through a rod is driven by a voltage difference and diffusion processes between the two contacts at $x = 0, L$. The hydrodynamic equations for electron and hole flow include Gauss's law in Eq. (5.21a), charge conservation equations for electrons and holes in Eqs. (5.21b) and (5.21c), and momentum conservation equations for electrons and holes in Eqs. (5.21d) and (5.21e). The changes of kinetic energy of electrons and holes due to their interactions with the lattice and the lattice thermal energy are described in Eqs. (5.21f), (5.21g), and (5.21h), respectively. The system of equations is

$$\frac{\partial^2 V}{\partial x^2} = -\frac{e}{\epsilon_r \epsilon_0} \left(n_h - n_e \right) \tag{5.21a}$$

$$\frac{\partial n_e}{\partial t} + \frac{\partial (u_e n_e)}{\partial x} = G_{e,net} \tag{5.21b}$$

$$\frac{\partial n_h}{\partial t} + \frac{\partial (u_h n_h)}{\partial x} = G_{h,net} \tag{5.21c}$$

$$\frac{\partial u_e}{\partial t} + u_e \frac{\partial u_e}{\partial x} = \frac{e}{m_e^\star} \frac{\partial V}{\partial x} - \frac{k_B}{m_e^\star n_e} \frac{\partial (n_e T_e)}{\partial x} - \frac{u_e}{\tau_{M,e}} \tag{5.21d}$$

$$\frac{\partial u_h}{\partial t} + u_h \frac{\partial u_h}{\partial x} = -\frac{e}{m_h^\star} \frac{\partial V}{\partial x} - \frac{k_B}{m_h^\star n_h} \frac{\partial (n_h T_h)}{\partial x} - \frac{u_h}{\tau_{M,h}} \tag{5.21e}$$

$$\frac{\partial T_e}{\partial t} + u_e \frac{\partial T_e}{\partial x} = -\frac{2}{3} T_e \frac{\partial u_e}{\partial x} + \frac{2}{3 n_e k_B} \frac{\partial}{\partial x} \left(k_e \frac{\partial T_e}{\partial x} \right) - \frac{T_e - T_L}{\tau_{E,e}}$$
$$+ \frac{2 m_e^\star u_e^2}{3 k_B \tau_{M,e}} \left(1 - \frac{\tau_{M,e}}{2 \tau_{E,e}} \right) \tag{5.21f}$$

$$\frac{\partial T_h}{\partial t} + u_h \frac{\partial T_h}{\partial x} = -\frac{2}{3} T_h \frac{\partial u_h}{\partial x} + \frac{2}{3 n_h k_B} \frac{\partial}{\partial x} \left(k_h \frac{\partial T_h}{\partial x} \right) - \frac{T_h - T_L}{\tau_{E,h}}$$
$$+ \frac{2 m_h^\star u_h^2}{3 k_B \tau_{M,h}} \left(1 - \frac{\tau_{M,h}}{2 \tau_{E,h}} \right) \tag{5.21g}$$

$$C_{L,v}\frac{\partial T_L}{\partial t} = \frac{\partial}{\partial x}\left(k_L\frac{\partial T_L}{\partial x}\right) + \frac{3n_e k_B}{2}\left(\frac{T_e - T_L}{\tau_{E,e}}\right)$$

$$+ \frac{3n_h k_B}{2}\left(\frac{T_h - T_L}{\tau_{E,h}}\right) + q_l^{*'''} \quad (5.21h)$$

where $V(x, t)$ is the electrostatic potential, t is the time, $n_e(x, t)$ is the electron density, $n_h(x, t)$ is the hole density, $u_e(x, t)$ is the electron velocity, $u_h(x, t)$ is the hole velocity, $T_e(x, t)$ is the electron temperature, $T_h(x, t)$ is the hole temperature, $T_L(x, t)$ is the lattice temperature, $q_l(x, t)$ represents the heat losses, $G_{e,net}(x, t) = G_e(x, t) - R_e(x, t)$ is the net generation rate of electrons which is equal to the generation rate $G_e(x, t)$ minus the recombination rate $R_e(x, t)$, and $G_{h,net}(x, t) = G_h(x, t) - R_h(x, t)$ is the net generation rates of holes, which is equal to the generation rate $G_h(x, t)$ minus the recombination rate $R_h(x, t)$.

The physical parameters are the elementary charge, e, the relative permittivity of the material, ϵ_r, the permittivity of free space, ϵ_0, the effective electron mass, $m_e^\star$, the effective hole mass, $m_h^\star$, Boltzmann's constant, k_B, the momentum relaxation time for electrons, $\tau_{M,e}$, the momentum relaxation time for holes, $\tau_{M,h}$, the energy relaxation time for electrons, $\tau_{E,e}$, the energy relaxation time for holes, $\tau_{E,h}$, the thermal conductivity of electrons, k_e, the thermal conductivity of holes, k_h, the thermal conductivity of the lattice, k_L, and the volumetric heat capacity of the lattice $C_{L,v}$.

Consider the case where electrons and holes are in thermal equilibrium, i.e. $T_e = T_h = T_c$, and their temperature is uniform in space and constant in time. Thermal equilibrium is relevant for devices where the densities of electrons and holes are similar in magnitude and when their mobilities are also similar in magnitude. The former criterion is true for intrinsic materials, for materials whose donor and acceptor doping levels are similar (not very common), and for transistor devices where the gate voltage would give rise to similar levels of "instantaneous doping," such as in graphene devices (Freitag et al., 2010, Bae et al., 2010, 2011). Meeting both criteria would mean that both of the carriers' speeds/kinetic energies are approximately the same. This is the case of intrinsic graphite, for which the ratio of the electron to the hole mobility is close to one (Klein, 1964) and for graphene transistors (Freitag et al., 2010, Bae et al., 2010, 2010). While thermal equilibrium is more physically feasible, keeping a constant and uniform carrier temperature may be a more crude approximation. It would mean that the external forces (e.g. electric field) are not large enough to change the carrier energy significantly in space or time.

Discarding the terms which involve a space or time derivative, and substituting the remaining terms from Eqs. (5.21f) and (5.21g) into the lattice energy equation, Eq. (5.21h), it becomes

$$
C_{L,v}\frac{\partial T_L}{\partial t} = \frac{\partial}{\partial x}\left(k_L\frac{\partial T_L}{\partial x}\right) - n_e k_B T_c\frac{\partial u_e}{\partial x} + \frac{n_e m_e^{\star} u_e^{2}}{\tau_{M,e}}\left(1 - \frac{\tau_{M,e}}{2\tau_{E,e}}\right)
$$

$$
- n_h k_B T_c\frac{\partial u_h}{\partial x} + \frac{n_h m_h^{\star} u_h^{2}}{\tau_{M,h}}\left(1 - \frac{\tau_{M,h}}{2\tau_{E,h}}\right) + q_l^{*\prime\prime\prime} \tag{5.22}
$$

Nondimensionalization of equations is a well-known technique for simplifying complex equations. Upon its discovery and development, this technique has been utilized historically and has proven fruitful in the development of science. This technique will be exploited here to significantly reduce the complexity of the hydrodynamic model.

Equations (5.21) along with Eq. (5.22) are nondimensionalized using the following parameter definitions: $x^* = x/L$, $V^* = V/V_0$, $n_e^* = n_e/N_0$, $n_h^* = n_h/N_0$, $u_e^* = u_e/U$, $u_h^* = u_h/U$, $T_L^* = T_L/T_0$, $T_c^* = T_c/T_0$, and $t^* = tU/L$, with V_0 a reference voltage, N_0 a reference density, and T_0 a reference temperature. The following relationship will be used in this step. The mobility is related to the momentum relaxation time as follows: $\mu_e = e\tau_{M,e}/m_e^{\star}$; this is the simplest relationship between these parameters. The nondimensional version of the equations becomes

$$
\frac{\partial^2 V^*}{\partial x^{*2}} = -\zeta\left(n_h^* - n_e^*\right) \tag{5.23a}
$$

$$
\frac{\partial n_e^*}{\partial t^*} + \frac{\partial (u_e^* n_e^*)}{\partial x^*} = G_{e,net}^* \tag{5.23b}
$$

$$
\frac{\partial n_h^*}{\partial t^*} + \frac{\partial (u_h^* n_h^*)}{\partial x^*} = G_{h,net}^* \tag{5.23c}
$$

$$
\text{Re}\,\frac{\tau_{M,e}}{\tau_{M,e,ref}}\left[\frac{\partial u_e^*}{\partial t^*} + u_e^*\frac{\partial u_e^*}{\partial x^*}\right] = \frac{\mu_e}{\mu_{e,ref}}\left[\frac{\partial V^*}{\partial x^*} - \beta\frac{1}{n_e^*}\frac{\partial n_e^*}{\partial x^*}\right] - u_e^* \tag{5.23d}
$$

$$
\text{Re}\,\frac{\tau_{M,e}}{\tau_{M,e,ref}}\left[\frac{\partial u_h^*}{\partial t^*} + u_h^*\frac{\partial u_h^*}{\partial x^*}\right] = -\frac{\mu_e}{\mu_{e,ref}}m_r\left[\frac{\partial V^*}{\partial x^*} + \beta\frac{1}{n_h^*}\frac{\partial n_h^*}{\partial x^*}\right] - \gamma u_h^*
$$

$$
\tag{5.23e}
$$

$$
\psi_0\frac{C_{L,v}}{C_{L,v,ref}}\frac{\partial T_L^*}{\partial t^*} = \psi_1\frac{\partial}{\partial x^*}\left(\frac{k_L}{k_{L,ref}}\frac{\partial T_L^*}{\partial x^*}\right)
$$

$$
- \beta\left(n_e^*\frac{\partial u_e^*}{\partial x^*} + m_r n_h^*\frac{\partial u_h^*}{\partial x^*}\right) + \left(1 - \frac{1}{2}v\right)
$$

$$
\times\left[\frac{\tau_{M,e,ref}}{\tau_{M,e}}n_e^* u_e^{*2} + \frac{\tau_{M,e,ref}}{\tau_{M,h}}n_h^* u_h^{*2}\right] + q_l^{*\prime\prime\prime}
$$

$$
\tag{5.23f}
$$

where $\zeta = eL^2 N_0 / \epsilon_r \epsilon_0 V_0$, $\beta = k_B T_c / eV_0$, $\gamma = \tau_{M,e} / \tau_{M,h}$, $m_r = m_e^\star / m_h^\star$, $v = \tau_{M,e} / \tau_{E,e} = \tau_{M,h} / \tau_{E,h}$, $\psi_0 = \tau_{M,e,ref} C_{L,v,ref} T_0 / m_e^\star N_0 UL$, $\psi_1 = \tau_{M,e,ref} k_{L,ref} T_0 / m_e^\star N_0 U^2 L^2$, $G_{e,net}^* = G_{e,net} L / N_0 U$, $G_{h,net}^* = G_{h,net} L / N_0 U$, $q_l^{*'''} = q_l''' \tau_{M,e} / N_0 m_e^\star U^2$, and $U = eV_0 \tau_{M,e,ref} / m_e^\star L = \mu_{e,ref} V_0 / L$ is the maximum average electron velocity. $\mathrm{Re} = U \tau_{M,e,ref} / L$, which is the Reynolds number for the electron cloud (Dyakonov and Shur, 1993, Lai and Majumdar, 1996, Mohseni et al., 2005, Osses-Márquez and Calderón-Muñoz, 2014). The Reynolds number here is analogous to that used in fluid dynamics; a small value means that the inertia forces in the momentum equations can be neglected (Dyakonov and Shur, 1993, Mohseni et al., 2005, Lai and Majumdar, 1996, Osses-Márquez and Calderón-Mu noz, 2014). Note that in the definition of v, it was assumed that the ratio of momentum relaxation time to energy relaxation time was equal for both electrons and holes.

Electrons and holes are expected to have significant scattering events when they are traveling through the rod; as a consequence, the Reynolds number becomes very small. Based on our nondimensionalized equations, the Reynolds number multiplies the transient and inertia forces in the momentum equations for electrons (5.23d) and holes (5.23e), which means that these terms can be safely discarded (Lai and Majumdar, 1996, Mohseni et al., 2005, Osses-Márquez and Calderón-Mu noz, 2014). Because the hydrodynamic model is composed of highly coupled nonlinear partial differential equations, such simplifications prove crucial in reducing its complexity.

Discarding the transient and inertia terms, the system of nondimensional equations (5.23) becomes

$$\frac{\partial^2 V^*}{\partial x^{*2}} = -\zeta \left(n_h^* - n_e^* \right) \tag{5.24a}$$

$$\frac{\partial (u_e^* n_e^*)}{\partial x^*} = G_{e,net}^* \tag{5.24b}$$

$$\frac{\partial (u_h^* n_h^*)}{\partial x^*} = G_{h,net}^* \tag{5.24c}$$

$$u_e^* = \frac{\mu_e}{\mu_{e,ref}} \left[\frac{\partial V^*}{\partial x^*} - \beta \frac{1}{n_e^*} \frac{\partial n_e^*}{\partial x^*} \right] \tag{5.24d}$$

$$u_h^* = -\frac{\mu_h}{\mu_{e,ref}} \left[\frac{\partial V^*}{\partial x^*} + \beta \frac{1}{n_h^*} \frac{\partial n_h^*}{\partial x^*} \right] \tag{5.24e}$$

$$0 = \psi_1 \frac{\partial}{\partial x^*} \left(\frac{k_L}{k_{L,ref}} \frac{\partial T_L^*}{\partial x^*} \right) - \beta \left(n_e^* \frac{\partial u_e^*}{\partial x^*} + m_r n_h^* \frac{\partial u_h^*}{\partial x^*} \right)$$
$$+ \left(1 - \frac{1}{2} v \right) \left[\frac{\tau_{M,e,ref}}{\tau_{M,e}} n_e^* u_e^{*2} + \frac{\tau_{M,e,ref}}{\tau_{M,h}} n_h^* u_h^{*2} \right] + q_l^{*'''} \tag{5.24f}$$

5.1.3.2 Drift-Diffusion Model

In this model, it is assumed that the carrier diffusion terms are small compared to the rest of the terms in Eqs. (5.24); the diffusion terms are those containing the coefficient β. This assumption is more appropriate for longer rods; as the length of the rod decreases, the importance (magnitude) of the diffusion terms needs to be assessed. Discarding the diffusion terms yields the following equations:

$$\frac{\partial^2 V^*}{\partial x^{*2}} = -\zeta \left(n_h^* - n_e^* \right) \tag{5.25a}$$

$$\frac{\partial (u_e^* n_e^*)}{\partial x^*} = G_{e,net}^* \tag{5.25b}$$

$$\frac{\partial (u_h^* n_h^*)}{\partial x^*} = G_{h,net}^* \tag{5.25c}$$

$$u_e^* = \frac{\mu_e}{\mu_{e,ref}} \frac{\partial V^*}{\partial x^*} \tag{5.25d}$$

$$u_h^* = -\frac{\mu_h}{\mu_{e,ref}} \frac{\partial V^*}{\partial x^*} \tag{5.25e}$$

$$0 = \psi_1 \frac{\partial}{\partial x^*} \left(\frac{k_L}{k_{L,ref}} \frac{\partial T_L^*}{\partial x^*} \right)$$
$$+ \left(1 - \frac{1}{2} v \right) \left[\frac{\tau_{M,e,ref}}{\tau_{M,e}} n_e^* u_e^{*2} + \frac{\tau_{M,e,ref}}{\tau_{M,h}} n_h^* u_h^{*2} \right] + q_l^{*\prime\prime\prime} \tag{5.25f}$$

The momentum and energy relaxation times depend on the energy of the carriers and the lattice (Jonscher, 1964, Bløtekjær, 1970, Wang, 1985, Majumdar et al., 1995a, Lundstrom, 2009, Osses-Márquez and Calderón-Munoz, 2014). In our definition of v, it has been assumed that $v = v_e = v_h$ to simplify the energy equation (5.25f). This definition is less restrictive than it appears to be. The only restriction is that each v, or ratio of relaxation times, be equal. Yet, note that (i) the momentum relaxation times for electrons and holes may have different values, and (ii) they may even have different energy dependencies.

It is known that the ratio of the electron to hole mobility in graphite is about 1.1 for low and near room temperatures (Klein, 1964), while in studies of graphene devices, the same value for both the electron and hole mobility has been used (Meric et al., 2008, Freitag et al., 2009, Bae et al., 2010). The momentum relaxation time can be readily calculated from the mobility. Though, values for the energy relaxation times have not been found in the literature. As a consequence, the effect of the energy relaxation time will be analyzed by changing the magnitude of v.

Let us have a short discussion about the implications of the relaxation times. The relaxation time is defined as the amount of time it takes a system to return to its equilibrium mode after it has been perturbed. The relaxation time is also the inverse of the frequency of collisions. In general, the momentum relaxation time is smaller than the energy relaxation time because if a carrier collision occurs, a change in momentum (magnitude or direction) is more likely to occur than a change in energy (magnitude only). Upon collision of a charge carrier with the lattice, the extent of energy transferred is determined by the energy relaxation time. A larger energy relaxation time means a lower amount of energy is exchanged between the charge carrier and the lattice. This can be easily seen through the third term on the right-hand side of Eqs. (5.21f) and (5.21g). However, as the energy relaxation time τ_E increases, the coefficient $(1 - \tau_M/\tau_E)$ of the fourth term in Eqs. (5.21f) and (5.21g) increases, and this term is really important because it represents Joule (self) heating (Bløtekjær, 1970). This coefficient represented by $(1 - v/2)$ in Eq. (5.25f), where v acts as a modulator of the Joule heating power. This coefficient ranges from a minimum of 0.5 (50%) Joule heating power when $v = 1$ to approximately (100%) Joule heating power when $(v \ll 1)$. For instance, if $\tau_E = 50\tau_M$, $v = 0.02$, and the coefficient $(1 - v/2) = 0.99$.

If we let the energy relaxation time τ_E be much greater than the momentum relaxation time τ_M, i.e. $(v \lesssim 0.02)$, the coefficient multiplying the bracketed term on Eq. (5.25f) becomes nearly equal to one. In addition, by substituting the velocity for both electrons and holes, Eqs. (5.25d) and (5.25e), once into the energy equation (5.25f), it can now be rewritten as follows:

$$0 = \psi_1 \frac{\partial}{\partial x^*} \left(\frac{k_L}{k_{L,ref}} \frac{\partial T_L^*}{\partial x^*} \right) + \left(n_e^* u_e^* - m_r n_h^* u_h^* \right) \frac{\partial V^*}{\partial x^*} + q_l^{*\prime\prime\prime} \qquad (5.26)$$

5.2 Physical Activation of Biochar with Nonthermal Plasma

There is a growing interest in using nonthermal plasma (NTP) to mitigate kinetic and thermodynamic obstacles to chemical conversions (Mehta et al., 2019). Processes that use NTP alone or in combination with a catalyst material to increase reaction rates, conversions, and yields have the potential to improve processes that can have great impact to society, such as NH_3 synthesis, CH_4 activation into hydrogen, or CO_2 conversion to value-added chemical and fuels (Bogaerts et al., 2020).

Steam activation of biochar utilizes a tremendous amount of energy since steam needs to be produced at temperatures of the order of 800 °C where the process can take from minutes to hours, depending on the characteristics of the feedstock. It has already been demonstrated that a microwave discharge can significantly reduce the apparent activation energy of the Boudouard reaction (Hunt et al., 2013):

$$CO_2(g) + C(s) \rightleftharpoons 2CO \tag{5.27}$$

Section 5.2.1 describes the results of performing steam activation of biochar in the presence of NTP but at steam temperatures around $400\,°C$ which is half of the typical conventional physical activation with steam.

5.2.1 Plasma-Steam Activation

Production of biochar from two different local biomass types, i.e. woodchips and peach pits, was accomplished by Phoenix Energy, utilizing a 500 kW for woodchips and a 1 MW biomass gasifier, for peach pits. These two types of feedstock are locally available in the San Joaquin Valley of California and were chosen because they represent forest and agricultural biomass available in this region.

The purpose of the study is to compare the surface area and porosity characteristics of commercially available activated carbon, and activated carbon produced from conventional steam activation as well as from NTP-enhanced steam activation.

5.2.1.1 Commercial Activated Carbon

A sample of commercial granular activated carbon (GAC) from coconut shells was obtained and analyzed for Brunauer, Emmett, and Teller (BET) surface area with nitrogen at 77 K. A surface area of $1000\,m^2/g$ with 95% microporosity was obtained.

5.2.1.2 Conventional Steam Activation

Conventional steam activation was accomplished for peach pits and ponderosa pine biochar. BET surface area was obtained for all samples produced and compared with the raw biochar obtained from the gasification process. Raw biochar had BET surface areas in the range between 1 and $20\,m^2/g$ with a low percentage (below 30%) of microporous surface. The very low surface area limits the capacity of biochar to be utilized as adsorption material in filters. Steam temperatures were varied between 750 and $850\,°C$ with steam flow rates between 0.8 and $4\,g/min$ and activation times between 15 and 45 minutes. A substantially larger BET surface area was obtained for the steam activated samples, which showed values in the range between 300 and $650\,m^2/g$. A BET surface area of $730\,m^2/g$ was obtained for ponderosa pine steam activated samples, with microporosity of the order of 30%. It is noted that burn off ranged between 30% and 60% of the original biochar sample mass.

5.2.1.3 Plasma-Enhanced Steam Activation

Conventional activation of biochar utilizing steam at temperatures around 800–1000 °C consumes a very large amounts of energy. Thus, the purpose of this experiment was to activate the biochar at a lower steam temperature but in the presence of nonthermal plasma. A nonthermal plasma reactor was designed and constructed, where the discharge was generated by inserting a steel coil inside the ceramic wall of the reactor. Nonthermal plasma was generated close to the surface of the coil and at ceramic walls, where the carbonaceous material located in the spaces between the coil becomes in contact with plasma and steam. Steam between 370 and 400 °C was used during the tests, where an input power of 100 W for the plasma discharge showed the best results for the tests performed. Although BET surface areas in the range of 300 m^2/g where obtained, which are low compared to commercial activated carbon, the micropore area obtained using plasma activation was of the order of 80–90% compared with a value around 30% of the conventional activation process. It is clear that plasma has a strong effect in increasing micropore area and a weak effect increasing BET surface area, although the latter one can be increased by raising steam temperature to values that are still lower than for conventional physical activation. In order to maximize the power provided for AC voltage, impedance matching is required.

References

N.R. Aluru, A. Raefsky, P.M. Pinsky, K.H. Law, R.J.G. Goossens, and R.W. Dutton. A finite element formulation for the hydrodynamic semiconductor device equations. *Computer Methods in Applied Mechanics and Engineering*, 107(1–2):269–298, 1993.

V.K. Arora, M.L.P. Tan, and C. Gupta. High-field transport in a graphene nanolayer. *Journal of Applied Physics*, 112(11):114330, 2012.

M.-H. Bae, Z.-Y. Ong, D. Estrada, and E. Pop. Imaging, simulation, and electrostatic control of power dissipation in graphene devices. *Nano Letters*, 10(12):4787–4793, 2010.

M.-H. Bae, S. Islam, V.E. Dorgan, and E. Pop. Scaling of high-field transport and localized heating in graphene transistors. *ACS Nano*, 5(10):7936–7944, 2011.

L. Ballestra, S. Micheletti, and R. Sacco. Semiconductor device simulation using a viscous-hydrodynamic model. *Computer Methods in Applied Mechanics and Engineering*, 191(47–48):5447–5466, 2002.

T.L. Bergman, F.P. Incropera, D.P. DeWitt, and A.S. Lavine. *Fundamentals of Heat and Mass Transfer*. John Wiley & Sons, 2011.

K. Bløtekjær. Transport equations for electrons in two-valley semiconductors. *IEEE Transactions on Electron Devices*, 17(1):38–47, 1970.

A. Bogaerts et al. The 2020 plasma catalysis roadmap. *Journal of Physics D: Applied Physics*, 53(443001):1–51, 2020.

M. Breusing, C. Ropers, and T. Elsaesser. Ultrafast carrier dynamics in graphite. *Physical Review Letters*, 102(8):086809, 2009.

W.R. Calderón-Muñoz. *Linear stability of electron-flow hydrodynamics in ungated semiconductors*. PhD thesis, University of Notre Dame, 2009.

R.K. Cook. Numerical simulation of hot-carrier transport in silicon bipolar transistors. *IEEE Transactions on Electron Devices*, 30(9):1103–1110, 1983.

V.E. Dorgan, M.-H. Bae, and E. Pop. Mobility and saturation velocity in graphene on SiO_2. *Applied Physics Letters*, 97(8):082112, 2010.

M. Dyakonov and M. Shur. Shallow water analogy for a ballistic field effect transistor: new mechanism of plasma wave generation by dc current. *Physical Review Letters*, 71(15):2465, 1993.

Entegris. Properties and characteristics of graphite. Technical report, Entegris, Inc., January 2015. http://poco.com/Portals/0/Literature/Semiconductor/IND-109441-0115.pdf.

M. Freitag, M. Steiner, Y. Martin, V. Perebeinos, Z. Chen, J.C. Tsang, and P. Avouris. Energy dissipation in graphene field-effect transistors. *Nano Letters*, 9(5):1883–1888, 2009.

M. Freitag, H.-Y. Chiu, M. Steiner, V. Perebeinos, and P. Avouris. Thermal infrared emission from biased graphene. *Nature Nanotechnology*, 5(7):497–501, 2010.

G. Fugallo, A. Cepellotti, L. Paulatto, M. Lazzeri, N. Marzari, and F. Mauri. Thermal conductivity of graphene and graphite: collective excitations and mean free paths. *Nano Letters*, 14(11):6109–6114, 2014.

R.O. Grondin, S.M. El-Ghazaly, and S. Goodnick. A review of global modeling of charge transport in semiconductors and full-wave electromagnetics. *IEEE Transactions on Microwave Theory and Techniques*, 47(6):817–829, 1999.

C.Y. Ho, R.W. Powell, and P.E. Liley. Thermal conductivity of the elements. *Journal of Physical and Chemical Reference Data*, 1(2):279–421, 1972.

X. Hong and D.D.L. Chung. Exfoliated graphite with relative dielectric constant reaching 360, obtained by exfoliation of acid-intercalated graphite flakes without subsequent removal of the residual acidity. *Carbon*, 91:1–10, 2015.

J. Hunt, A. Ferrari, A. Lita, M. Crosswhite, B. Ashley, and A.E. Stiegman. Microwave-specific enhancement of the carbon–carbon dioxide (boudouard) reaction. *Journal of Physical Chemistry C*, 117:26871–26880, 2013.

A.K. Jonscher. Transport of hot injected plasmas in semiconductors. *Proceedings of the Physical Society*, 84(5):767, 1964.

C.A. Klein. Pyrolytic graphites: their description as semimetallic molecular solids. *Journal of Applied Physics*, 33(11):3338–3357, 1962.

C.A. Klein. STB model and transport properties of pyrolytic graphites. *Journal of Applied Physics*, 35(10):2947–2957, 1964.

J. Lai and A. Majumdar. Concurrent thermal and electrical modeling of sub-micrometer silicon devices. *Journal of Applied Physics*, 79(9):7353–7361, 1996.

M. Lundstrom. *Fundamentals of Carrier Transport*. Cambridge University Press, 2009.

A. Majumdar, K. Fushinobu, and K. Hijikata. Effect of gate voltage on hot-electron and hot phonon interaction and transport in a submicrometer transistor. *Journal of Applied Physics*, 77(12):6686–6694, 1995a.

A. Majumdar, K. Fushinobu, and K. Hijikata. Heat generation and transport in submicron semiconductor devices. *Journal of Heat Transfer*, 117:25–31, 1995b.

J.W. McClure. Band structure of graphite and de Haas-van alphen effect. *Physical Review*, 108(3):612, 1957.

P. Mehta, P. Barboun, D.B. Go, J.C. Hicks, and W.F. Schneider. Catalysis enabled by plasma activation ofstrong chemical bonds: a review. *ACS Energy Letters*, 4:1115–1133, 2019.

I. Meric, M.Y. Han, A.F. Young, B. Ozyilmaz, P. Kim, and K.L. Shepard. Current saturation in zero-bandgap, top-gated graphene field-effect transistors. *Nature Nanotechnology*, 3(11):654, 2008.

K. Mohseni, A. Shakouri, R.J. Ram, and M.C. Abraham. Electron vortices in semiconductors devices a. *Physics of Fluids*, 17(10):100602, 2005.

A. Muñoz-Hernández. *Charge and Joule Heat Transport in Carbonaceous Materials and Activation of Biochar*. PhD dissertation, University of California - Merced, 2018.

A.H.C. Neto, F. Guinea, N.M.R. Peres, K.S. Novoselov, and A.K. Geim. The electronic properties of graphene. *Reviews of Modern Physics*, 81(1):109, 2009.

K.S. Novoselov, A.K. Geim, S.V. Morozov, D.A. Jiang, Y. Zhang, S.V. Dubonos, I.V. Grigorieva, and A.A. Firsov. Electric field effect in atomically thin carbon films. *Science*, 306(5696):666–669, 2004.

J.I. Osses-Márquez and W.R. Calderón-Mu noz. Thermal influence on charge carrier transport in solar cells based on GaAs PN junctions. *Journal of Applied Physics*, 116(15):154502, 2014.

A.W. Smith and K.F. Brennan. Hydrodynamic simulation of semiconductor devices. *Progress in Quantum Electronics*, 21(4):293–360, 1998.

S.M. Sze and K.K. Ng. *Physics of Semiconductor Devices*. John Wiley & Sons, 2006.

I.N. Volovichev, J.E. Velazquez-Perez, and Yu.G. Gurevich. Transport boundary conditions for solar cells. *Solar Energy Materials and Solar Cells*, 93(1):6–10, 2009.

P.R. Wallace. The band theory of graphite. *Physical Review*, 71(9):622, 1947.

C.T. Wang. A new set of semiconductor equations for computer simulation of submicron devices. *Solid-State Electronics*, 28(8):783–788, 1985.

6

Numerical Simulations

6.1 Background

With the ever increasing computational power and the relatively easy access to parallel computing architectures that reduce computational processing times to a fraction of what it took to run in older computers, numerical simulations of physical processes are being able to model significantly more complex physical phenomena. Research in plasma discharges has benefited tremendously from these advanced simulations by being able to model two and three dimensional geometries, as well as processes that include multiphysics analysis, incorporating reactive species with rates that cover time scales that differ by several orders of magnitude.

From the simple mathematical model of thermal arcs using the Elenbaas-Heller equation, shown in Section 3.1.1, recent progress in computational analysis of DC thermal torches and plasma spray devices has contributed to a better understanding of the dynamics of the electric arc and its reattachment (Trelles et al., 2006, 2009). Moreover, modeling of thermal plasmas has predominantly utilized the local thermodynamic equilibrium (LTE) assumption in which electrons, ions, and neutral gas depend on the local thermodynamic state, i.e. a single value of temperature and pressure. However, it has been realized that thermodynamic nonequilibrium in thermal plasmas is particularly important in regions where plasma and gas interact (Chen et al., 1981, Chazelas et al., 2017).

These nonequilibrium conditions, which are common in nonthermal plasmas but not typically found in computational fluids, heat, and mass transfer, will be the main focus of this chapter.

6.2 Modeling Approaches

Plasmas can be considered as a gas with charged particles. In general, electrons, ions, and neutral species are the main components of a plasma discharge, however, negative ions, dust particles, liquid droplets, and other components can also be found in complex plasma flows (Brieda, 2019). In addition, physical phenomena such as turbulence, radiation, magnetic reconnection, and instabilities can occur in plasma flows. This chapter will concentrate on the main aspects of plasma modeling leaving more complex applications to other published works found in the literature. Two of the most common types of plasma simulations are the kinetic and fluid approaches.

6.2.1 Kinetic Approach

In the kinetic approach, the positions and velocities of individual particles are tracked in a Lagrangian framework. Using the leap-frog integration method, the two governing equations are integrated separately (Birdsall and Langdon, 2004).

$$m\frac{d\mathbf{v}}{dt} = \mathbf{F} \tag{6.1}$$

$$\frac{d\mathbf{x}}{dt} = \mathbf{v} \tag{6.2}$$

where $\mathbf{x}$ is the position, $\mathbf{v}$ is the velocity, and $\mathbf{F}$ is the force. These equations can be numerically integrated with respect to time using finite-differences approximations:

$$m\frac{\mathbf{v}_{new} - \mathbf{v}_{old}}{\Delta t} = \mathbf{F}_{old} \tag{6.3}$$

$$\frac{\mathbf{x}_{new} - \mathbf{x}_{old}}{\Delta t} = \mathbf{v}_{new} \tag{6.4}$$

Neglecting magnetic effects, the force generated by the electric field is

$$\mathbf{F} = q\mathbf{E} \tag{6.5}$$

where q is the charge, $\mathbf{E}$ is the electric field calculated at the particle, and where $\mathbf{E}$ can be calculated as the negative of the gradient of the electric potential ϕ.

$$\mathbf{E} = -\nabla\phi \tag{6.6}$$

The electric field is related to the charge density ρ by the Poisson's equation:

$$\nabla \cdot \mathbf{E} = \frac{\rho}{\epsilon_0} \tag{6.7}$$

where $\epsilon_0 \approx 8.8542 \times 10^{-12}$ C/(V m) is the permittivity of free space. Combining both equations and using finite differences approximation for a 1D case, the electric potential distribution can be obtained as:

$$\frac{\phi_{j-1} - 2\phi_j + \phi_{j+1}}{(\Delta x)^2} = -\frac{\rho_j}{\epsilon_0} \tag{6.8}$$

where j is the index for the finite difference discretization which denotes the location along the grid, which for a uniform grid has values from 0 to $L/\Delta x$, where L is the length of the domain and Δx is the distance between nodes. By considering boundary conditions for the domain, the system of equations can be solved directly. Periodic boundary conditions allows the use of the fast Fourier transform to obtain $\rho(\mathbf{k})$ and $\phi(\mathbf{k})$, where $\mathbf{k}$ is the wavenumber. These functions can be transformed back to $\rho(\mathbf{x})$ and $\phi(\mathbf{x})$ (Birdsall and Langdon, 2004) and the particle location and its velocity can be obtained using Eqs. (6.3) and (6.4).

Nonthermal plasmas produced inside vacuum chambers usually operate at pressures of the order of 10^{-6} Torr ($\approx 1.3 \times 10^{-4}$ Pa). Using ideal gas law, the number of particles in one cubic centimeter is 3.3×10^{10}. For a simulation of one cubic centimeter and considering integration on time for around ten thousand steps, the simulation needs to keep track of all these particles location, velocities, electric field, current densities, and forces during the ten thousand steps. This is a very inefficient process not only due to the computational time but also for the computer memory storage capacity. Many methods have been proposed that utilize the velocity distribution function (VDF) to reduce the number of calculations needed. Among these, particle-in-cell, Vlasov solvers, and Monte Carlo methods have significantly increased the efficiency of calculations of kinetic solvers (Brieda, 2019, Tajima, 2004).

Despite these improvements, nonthermal plasma discharges for energy systems tend to operate at atmospheric pressures, which means that number densities are of the order of $n = 2.7 \times 10^{19}$ molecules per cubic centimeter. Higher number densities imply that the distance that a molecule travels before it collides with another, known as the *mean free path*, is shorter, as it scales with the inverse of the number density:

$$\lambda = \frac{1}{\sigma n} \tag{6.9}$$

where σ is the *collision cross section*. For gases at very low pressures, the mean free path might approximate the dimensions of the chamber (L) and in some cases, it might be larger than L. In these type of cases, it becomes necessary to analyze plasmas from the kinetic point of view. As pressure is increased and the number density becomes higher, the mean free path becomes much smaller than the chamber dimensions. This ratio of quantities is given by the dimensionless parameter referred to as the *Knudsen number.*

$$Kn = \frac{\lambda}{L} \tag{6.10}$$

Thus, if Kn is much larger than one, $\lambda \gg L$ and collisions with the walls are more frequent than collisions between molecules, so particle collisions can be ignored. When $Kn \approx 1$ then collisions between particles are still infrequent and kinetic analysis is still needed. However, if $Kn \ll 1$, then the mean free path is much smaller than the dimensions of the chamber and the number density is large. In this case, we can regard the flow as being composed of so many particles that the fluid behaves as a continuum (Tajima, 2004). Under these conditions, the analysis of plasma discharges can be performed using the fluid model approach.

6.2.2 Fluid Model Approach

Consider a volume of gas composed of a large number of molecules that is moving in a certain space. We can expect that some molecules will have higher velocities than others and therefore, collisions will occur between adjacent molecules, which will slow down molecules that were moving fast and speed up molecules that were moving slow. As this process occurs, we expect that the transfer of momentum due to collisions will be eventually reduced to zero as equilibrium is reached. A distribution can be generated that shows the number of molecules in the gas traveling within a certain range of velocities. These velocities can be considered in different directions, so the relation between speed and velocities is given by the expression $v^2 = v_x^2 + v_y^2 + v_z^2$. Once velocities and energies are in equilibrium, the function that describes this state is called the Maxwell–Boltzmann distribution, given by the expression:

$$f(v) = \frac{dn_v}{dv} = \frac{4n}{\pi^{1/2}} \left(\frac{m}{2kT} \right)^{3/2} v^2 \exp\left[-\frac{mv^2}{2kT} \right] \tag{6.11}$$

It is important to note that by integrating Eq. (6.11) over the velocity moments for $0 \leq v \leq \infty$, the following quantities are obtained:

$\int_0^\infty f \, dv = n$ (gas number density)

$\frac{1}{n} \int_0^\infty v f \, dv = \left(\frac{8kT}{\pi m} \right)^{1/2} = \bar{v}$ (mean thermal velocity)

Defining kinetic energy as:

$$w = \frac{1}{2} mv^2 \tag{6.12}$$

the distribution of energies between w and $w + dw$ is:

$$f(w) = \frac{dn_w}{dv} = \frac{2n}{\pi^{1/2}} \frac{w^{1/2}}{(kT)^{3/2}} \exp\left[-\frac{w}{kT}\right] \tag{6.13}$$

and the mean energy of a gas with a Maxwellian distribution becomes:

$$\overline{w} = \frac{1}{n} \int_0^\infty w f(w) dw = \frac{3}{2} kT \tag{6.14}$$

More details of these derivations are found in Roth (1995). As mentioned above, energy and biomass processing applications involving plasmas tend to operate near or above atmospheric conditions. In addition, due to the need of processing large quantities of material (forest wood, agricultural waste, etc.), the dimensions of the reactors are fairly large. Therefore, the mean free path of molecules in an atmospheric-pressure gas is much smaller than the characteristic dimensions of the reactors. Thus, the Knudsen number is very small and the continuum approximation is valid. Under the continuum approximation, the modeling of gases and plasmas can be performed by using the Boltzmann equation.

$$\frac{\partial f}{\partial t} + \mathbf{v} \cdot \nabla f + \mathbf{F} \cdot \nabla_v f/m = \left(\frac{\partial f}{\partial t}\right)_{col} \tag{6.15}$$

where f is the distribution function, $\mathbf{v}$ is the charged particle velocity, $\mathbf{F}$ is the force, m is the mass, and ∇_v is the operator in the velocity space. The term on the right-hand side of the equation corresponds to the changes in the distribution due to particle collisions (Gogolides and Sawin, 1992). As similarly done by integrating Eq. (6.11) with respect to its velocity moments, Eq. (6.15) is integrated for its first three moments, which correspond to integrals with zeroth, first, and second power of the velocity, to give the mass, momentum, and energy balances. The governing equations take the form (Bogaerts et al., 2010, Farouk et al., 2006):

Continuity:

$$\frac{\partial n}{\partial t} + \nabla \cdot \mathbf{j} = R_{prod} - R_{loss} \tag{6.16}$$

$$\mathbf{j} = \mu n \mathbf{E} - D\nabla n \tag{6.17}$$

where R_{prod} and R_{loss} are the reaction rates of production and loss, respectively. Some of the other quantities present in the equations will be described in Section 6.3.1.

Momentum equation:

$$\frac{\partial n\mathbf{v}}{\partial t} + \nabla \cdot (n\mathbf{v}\mathbf{v}) = -\nabla p + \nabla \cdot \sigma + \sum_k n_k \mathbf{F}_k \tag{6.18}$$

$$\nabla \cdot \sigma = \mu \nabla^2 \mathbf{v} + \frac{\mu}{3} \nabla(\nabla \cdot \mathbf{v}) \tag{6.19}$$

Energy balance:

$$\frac{\partial w}{\partial t} + \nabla \cdot \mathbf{q} = -e\mathbf{j} \cdot \mathbf{E} + R_{w,loss} \tag{6.20}$$

$$\mathbf{q} = \frac{3}{5}\mu w \mathbf{E} - \frac{5}{3}D\nabla w_e \tag{6.21}$$

It is noted that this process has to be performed for each type of particle being analyzed, i.e. electrons, positive ions, negative ions, and neutral gas. Sections 6.3.1–6.3.3 provide some examples of simulations using the fluid approach.

6.3 Examples of Nonthermal Plasma Modeling

As mentioned in Section 6.2.2, the nonequilibrium characteristics of cold plasmas require determining quantities such as number densities, velocities, and energy for different species such as electrons, ions, and neutral molecules. In many applications, the ion and neutral gas temperature remain near ambient condition and thus, many formulations only account for the mean energy of the electrons, where the electron temperature can be obtained using Eq. (6.14) for a Maxwellian distribution. A few examples of numerical solutions from the literature that provide important insights about nonequilibrium plasma discharges are shown in in Sections 6.3.1–6.3.3. The purpose is not to replicate all the details of the simulations, but to direct the attention of the reader to specific aspects of nonthermal plasma simulations that are not found in typical simulations of fluids, heat, and mass transfer.

6.3.1 Cathode Fall of a DC Glow Discharge

In DC glow discharges, the cathode fall is the region where the electric potential drop occurs near the cathode, and is followed by a quasineutral region. The length of this region is usually characterized by a distance d_c. Well established models have been described in the literature that cover low-pressure operation (Ward, 1958, 1962, Raizer, 1991), atmospheric-pressure plasma (Shi and Kong, 2003), and field-emission driven microplasmas (Venkattraman, 2013).

Depending on the type of application, the governing equations for nonthermal plasma for the electrons, positive and negative ions, excited species, etc., have to be considered depending on the importance of each species in the overall behavior of the discharge. An analysis of a helium DC glow discharge where only electrons (subscript e) and positive ions (subscript i) are studied is described as follows. The one-dimensional electron and ion continuity, Poisson equation, current density, and drift velocity equations are given as follows:

$$\frac{\partial n_e}{\partial t} + \frac{\partial (n_e v_e)}{\partial z} = \alpha n_e v_e - R n_e n_i \tag{6.22}$$

$$\frac{\partial n_i}{\partial t} + \frac{\partial (n_i v_i)}{\partial z} = \alpha n_e v_e - R n_e n_i \tag{6.23}$$

$$\frac{\partial \epsilon E}{\partial z} = | e | (n_i - n_e) \tag{6.24}$$

$$J_{i,e} = \pm | e | n_{i,e} v_{i,e} \tag{6.25}$$

$$v_{i,e} = \pm \mu_{i,e} E - \frac{D_{i,e}}{n_{i,e}} \frac{\partial n_{i,e}}{\partial z} \tag{6.26}$$

where n_e is the number density of electrons and n_i the number density of ions, $J_{i,e}$ and $v_{i,e}$ are the current densities and drift velocities of ions and electrons, $\mu_{i,e}$ and $D_{i,e}$ are the mobilities and diffusion coefficients, respectively, α is the ionization coefficient, and R is the recombination coefficient. For a DC discharge, the transient terms can be neglected. In addition, for helium, the recombination rate is much smaller than the ionization rate. Considering $J = J_i + J_e$ and $j_{i,e} = J_{i,e}/J$, the governing equations can be reduced to:

$$\frac{dj_e}{dz} = \alpha j_e \tag{6.27}$$

$$\frac{dE}{dz} = -\frac{J}{\epsilon_0 \epsilon_r v_i} \left[1 - \left(1 + \frac{v_i}{v_e} \right) j_e \right] \tag{6.28}$$

Equation (6.26) can be added to this system of equations; however, measurements of drift velocities are available for several gases (Ward, 1962, 1958). In plasma discharges, several variables such as number densities, velocities, and electric fields can vary by several orders of magnitude. This fact results in numerical discretizations in space and time having to use time steps and distances between nodes that are very small, impacting the computational time needed to solve the equations. Therefore, it is recommended in many cases to convert the governing equations to dimensionless variables. Adopting the following dimensionless variables, $z^* = z/L$, $v^* = v/\mathscr{V}$, $E^* = E/\mathscr{E}$, where $L = 0.981$ cm, $\mathscr{V} = 10^8$ cm/s, and $\mathscr{E} = 14{,}000$ V/cm, the governing equations are converted to a dimensionless form.

$$\frac{dj_e}{dz^*} = \alpha L j_e \tag{6.29}$$

$$\frac{dE^*}{dz^*} = -\frac{\kappa}{v_i^*}\left[1 - \left(1 + \frac{v_i^*}{v_e^*}\right)j_e\right]$$ (6.30)

where $\kappa = JL/(\epsilon_0\epsilon_r \mathcal{E} \mathcal{V})$. The system of equation is solved using a fourth-order Runge–Kutta integrating scheme by guessing the value of the electric field at the cathode for a fixed value of the J. If $j_e > 1$ or $E^* < 0$ during the integration, then a new value of $E^*(0)$ is guessed and the integration along the discharge length z^* is repeated. Ward (1962), compiled a wide range of values of current densities for discharges of different gases. For instance for helium the current density divided by the square of the pressure remains in the range: $\frac{J}{p^2} \in [2,5]$ μA/cm^2 Torr^2, and the pressure times the normal cathode fall distance $pd_c \in [1.3, 1.45]$ Torr cm. In addition, considering a secondary Townsend coefficient of $\gamma = 0.25$, which represents the average number of electrons emitted at the cathode due to the impact of one ion, the normalized electron current density j_e has a value of 0.2 at the cathode ($z^* = 0$), while its value at the anode is $j_e = 1$. In order to solve the system of equations, the value of the first Townsend ionization coefficient (α) needs to be known. A first approximation was used by Ward (1962) assuming that the electrons at the cathode fall were in equilibrium with the electric field. Thus, the value of α was obtained as a function of the ratio E/p (E along the length of the discharge), which adopts the form:

$$\alpha = Ap \exp\left[-B\left(\frac{p}{E}\right)^{1/2}\right]$$ (6.31)

where the values of the constants are $A = 6.5$ (cm Torr)$^{-1}$ and $B = 16.4$ V$^{1/2}$/ (cm Torr)$^{1/2}$ for helium. The calculation of the ionization coefficient using only E/p, overestimates its magnitude near the cathode, so to improve the model, researchers have used the equation for the mean electron energy ($\bar{\epsilon}$):

$$\frac{d\bar{\epsilon}}{dz} = E(z) - (\bar{\epsilon} + U_i)\alpha - \frac{1}{2}U_e\alpha$$ (6.32)

where $U_i = 24.54$ (eV) is the onset potential for ionized states, and $U_e = 21.45$ (eV) is the onset potential of excited states for helium (An et al., 1977, Shi and Kong, 2003). Figure 6.1 shows the results of the dimensionless quantities: electron current density, electric field, and electron and ion densities for a DC discharge at $p = 100$ Torr (≈ 13 kPa), where the weighting function proposed by Shi and Kong (2003) has been used for calculating α. Number densities for the electrons and ions have been scaled by 10^{11} cm^{-3}. The two number densities (electrons and ions) differ significantly near the cathode as electrons are repelled and ions are attracted, but they are almost the same for the rest of the discharge, where a small difference is observed near the anode.

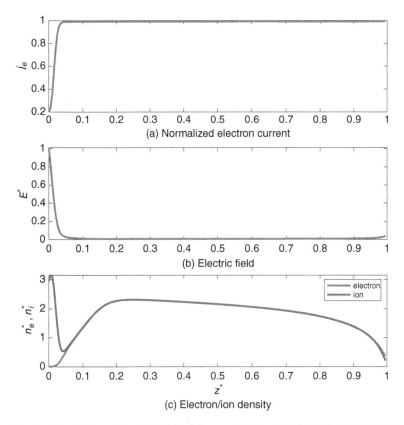

Figure 6.1 Mathematical model for DC glow discharge. (a) j_e, (b) E^*, and (c) n_e^*, n_i^*.

Figure 6.2 shows the values of the magnitude of the dimensionless electron and ion velocities where it is observed that the ion velocity is orders of magnitude slower than the electron velocity.

6.3.2 RF Plasma Discharge

Radio frequency (RF) plasma discharges are found in processes related to electrical circuit fabrication, pollutant decomposition, ozone generation, and many other applications. The key aspect is the oscillating voltage provided by the power supply, which exposes fundamental characteristics of nonequilibrium discharges. As electrons are much lighter than ions, the oscillatory nature of the input voltage implies that electrons move very fast from one electrode to the other, while the slower ions are not able to reach the electrodes and remain somewhere near the middle of the discharge gap.

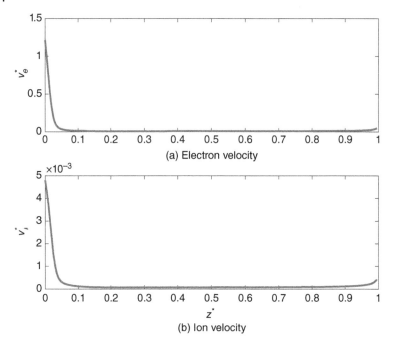

Figure 6.2 Mathematical model for DC glow discharge. (a) v_e^*, (b) v_i^*.

Gas pressure and frequency of oscillation will affect the extent of the displacement of electrons and ions.

The formulation of the governing equations given in Section 6.3.1 can be modified and written in slightly different ways. For instance, the source term in the continuity equation can be given in a Arrhenius exponential form as in Nitschke and Graves (1994), and the drift velocity given by Eq. (6.26) can be modified using the Einstein's relation $D = \mu \frac{k_b T}{q}$. Following the formulation by Hammond et al. (2002), the nondimensional electron and ion continuity equations and electron number density flux are given as:

$$\frac{\partial n_e}{\partial t} + \nabla \cdot \mathbf{j}_e = p_1 n_e e^{-p_2/T_e} \tag{6.33}$$

$$\frac{\partial n_i}{\partial t} + \nabla \cdot (n_i \mathbf{v}_i) = p_1 n_e e^{-p_2/T_e} \tag{6.34}$$

$$\mathbf{j}_e = -p_7 \left(n_e \mathbf{E} + \nabla(n_e T_e) \right) \tag{6.35}$$

where p_1, p_2, and p_7 are dimensionless constants (Hammond et al., 2002). The cathode fall discharge formulation shown in the previous example used correlations from measured drift velocities for the ions in a helium discharge. Instead, for the RF discharge a momentum balance can be used to solve for the ion velocity.

$$\frac{\partial \mathbf{v_i}}{\partial t} + \mathbf{v_i} \cdot \nabla \mathbf{v_i} = p_3 \mathbf{E} - p_4 |\mathbf{v_i}| \mathbf{v_i} \tag{6.36}$$

Finally, the mean electron energy equation given by Eq. (6.32) can be replaced by the thermal energy balance equation, where the ion energy equation is neglected.

$$\frac{\partial (\frac{3}{2} n_e T_e)}{\partial t} + \nabla \cdot \mathbf{q_e} = -\mathbf{j_e} \cdot \mathbf{E_e} - p_5 n_e e^{-p_2/T_e} - p_6 n_e (T_e - T_{neut}) \tag{6.37}$$

$$\mathbf{q_e} = \frac{5}{2} \mathbf{j_e} T_e - p_8 n_e T_e \nabla T_e \tag{6.38}$$

where $\mathbf{q_e}$ is the electron thermal energy flux, and the right-hand side terms denote Joule heating, energy loss per ionizing collision, and energy loss per electron-neutral collisions, respectively Nitschke and Graves (1994). The values of the dimensionless parameters p_i for $i \in [1,8]$ and the details of the numerical implementation are described in Hammond et al. (2002). However, an important point in this solution is related to Eq. (6.35). As electrons oscillate from side to side, at certain points of the solution, the term $n_e \mathbf{E}$ is dominant and at other points, the term $\nabla (n_e T_e)$ dominates. The equations for n_e and $n_e T_e$ are very stiff so it is recommended to use implicit time advancement schemes while the ion equations can be solved with explicit solvers. Moreover, the discretization using a staggered grid can be significantly improved based on the formulation used. For instance, for the one-dimensional case, the use of the central-difference formulation in Eq. (6.35) gives:

$$j_{e,k+1/2} = -p_7 \left[\left(\frac{n_{e,k} + n_{e,k+1}}{2} \right) E_{k+1/2} + \frac{n_e T_{e,k+1} - n_e T_{e,k}}{\Delta x_k} \right] \tag{6.39}$$

requires a very fine grid since both n_e and $n_e T_e$ change by several orders of magnitude near the electrodes, and this increases significantly the computational time. Due to this constraint, a discretization that is used very often is the Scharfetter–Gummel formulation (Scharfetter and Gummel, 1969); however, this scheme overpredicts the plasma density by 20%. Hammond et al. (2002) introduced the concept of mean velocity. In one dimension, Eq. (6.35) can be written as:

$$j_e = -p_7 n_e \left(\mathbf{E} + \frac{1}{n_e} T_e \frac{\partial n_e}{\partial x} + \frac{\partial T_e}{\partial x} \right) \tag{6.40}$$

which is equivalent to

$$j_e = -p_7 n_e \left(\mathbf{E} + T_e \frac{\partial \ln(n_e)}{\partial x} + \frac{\partial T_e}{\partial x} \right) = n_e v_e \tag{6.41}$$

The mean velocity is then defined as (Hammond et al., 2002):

$$\bar{v}_{e,k+1/2} = -P_7 \left[-\frac{\Phi_{k+1} - \Phi_k}{\Delta x_k} + \frac{T_{e,k+1/2}}{\Delta x_k} \ln \frac{n_{e,k+1}}{n_{e,k}} + \frac{T_{e,k+1} - T_{e,k}}{\Delta x_k} \right]$$

(6.42)

where Φ is the electric potential. This formulation removes the need to calculate discrete derivatives containing n_e and $n_e T_e$ and instead requires the computation of $\ln \left(\frac{n_{e,k+1}}{n_{e,k}} \right)$, which behaves in a much smoother way even for coarser grids. Results show differences of less than 1% in the peak electron number density with different number of grid points (Hammond et al., 2002). Figure 6.3 shows results of the numerical model for different variables related to the RF plasma discharge for a pressure of 250 mTorr ($\approx$ 33 Pa) for a helium discharge with 100 uniform subdivisions for a gap of 0.04 m. The frequency of oscillation is 12 MHz with a sinusoidal voltage input between $\pm$500 V. The figure shows the values (with dimensions) of the variables after 500 RF cycles. Figure 6.3a depicts the ion density (n_i) that clearly shows how the slow ions tend to concentrate in the middle of the discharge gap. On the other hand, Figure 6.3b,c shows that at the start of a new RF cycle, the voltage is $\Phi = -500$ V at the left electrode and therefore the electrons have been repelled showing a number density (n_e) of the order of 10^{-40} m^{-3} at the left electrode. This very large variation in the magnitude of n_e is the reason why central difference schemes for the drift diffusion equation require a very high grid density and schemes such as the ones presented in Hammond et al. (2002) or Scharfetter and Gummel (1969) are needed. Figure 6.3d shows the mean electron energy which is around 6.5 eV at the left electrode and decreases to around 5.3 eV at the right electrode. Figure 6.4a shows the variation of the electric field E along the discharge gap. It is observed that it has a value of approximately -12 kV/m at the left electrode which increases until $x = 0.01$ m and then remains almost constant. It is important to indicate that the magnitude of these variables varies throughout the RF cycle, but here we are showing at this specific condition. Figure 6.4b shows the ion velocity v_i calculated using Eq. (6.36). It shows how ions have a low velocity toward the center of the gap. Finally, the logarithmic axes of Figure 6.3a,b provide a good description of the difference in the order of magnitude between the ion and electron densities at the left electrode, but it is hard to verify that quasineutrality is observed in the rest of the discharge. Figure 6.4c shows a linear y axis which allows to show that in the sheath near the left electrode, the number densities differ by orders of magnitude, but for values of $x > 0.01$ m, the values of the ion and electron densities coincide.

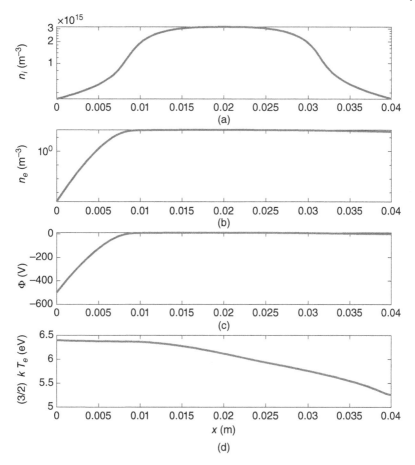

Figure 6.3 Numerical model for RF glow discharge. (a) n_i, (b) n_e, (c) Φ, (d) $(3/2)kT_e$.

6.3.3 Plasma Chemistry

The two previous examples taken from the literature provide details about specific aspects of the behavior of nonthermal plasma discharges under DC and RF input voltages. For instance, it is observed how the electric field becomes affected by the presence of the charged particles, the number densities of electrons and ions differ near the electrodes, and the quasineutral nature of plasma is observed in the hydrodynamic section of the model. Graphic representations of the large variation of the electron number density near the cathode were also shown, which is part of the reason why formulations such as the discretization method of (Scharfetter

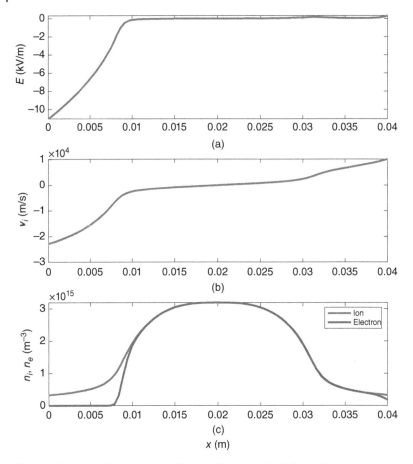

Figure 6.4 Numerical model for RF glow discharge. (a) E, (b) v_i, (c) n_i, n_e.

and Gummel, 1969) or the mean electron velocity definition by Hammond et al. (2002) have been proposed to speed-up numerical simulation of these discharges. However, the two examples were developed for discharges in helium and at pressures below atmospheric conditions, for which different characteristics of the behavior of electrons and He$^+$ ions were studied. Due to the type of applications in the semiconductor industry, surface cleaning, plasma-enhanced chemical vapor deposition, etc., a large number of published papers analyzes discharges in helium, argon, oxygen, or nitrogen. On the other hand, energy-related applications, especially for the conversion of biomass or biochar as well as the treatment of product gases generated during these processes, tend to operate at atmospheric or higher pressures and with gases such as air, pyrolysis or gasification product gas, tars, synthesis gas, or mixtures of pollutants (H_2S, SO_2, etc.). Therefore, not only the number of species present grows significantly, but also the number of possible reactions can easily be in the hundreds or even thousands. If we add the fact the electrical discharges will also produce excited species, ionized species and radicals, most of which have reaction times that are orders of magnitude shorter than equilibrium-chemistry reactions, then we can see why the numerical simulation of these applications is a very challenging topic that is still a subject of intense research. The detailed representation of the plasma chemistry in applications such as tar decomposition of a gasifier product gas, or the removal of nitrogen oxides (NO_x) from combustion exhaust streams using plasma discharges, is beyond the scope of this book. However, the next example intends to provide an insight of the way that chemical reactions are included in the simulation of nonthermal plasma discharges. The main steps used in the formulation used by Farouk et al. (2006) for the simulation of an atmospheric-pressure DC micro-glow discharge in argon are described in this section. The kinetic processes considered are shown in Table 6.1, where the rate coefficients can be obtained from the literature. As usual, the continuity equation for the electrons is given as:

$$\frac{\partial n_e}{\partial t} + \nabla \cdot \mathbf{j_e} = \dot{n}_e \tag{6.43}$$

$$\mathbf{j_e} = \mu_e n_e \nabla \phi - D_e \nabla n_e \tag{6.44}$$

A continuity equation for each of the ions, electronically excited, and neutral species has the form:

$$\frac{\partial n_i}{\partial t} + \nabla \cdot (n_i \mathbf{v}_i) = \dot{n}_i \tag{6.45}$$

where the index i corresponds to species: Ar^+, Ar_2^+, Ar^*, and Ar. The momentum equations are given as Eq. (6.18) for each of the positive ions, excited atoms, and neutral atoms, where p is the sum of the partial pressures of these particles, and the viscous stress tensors are given by Eq. (6.19) for each of these species. In general, ions and neutral species remain at temperatures near ambient, so the energy balance is usually performed only for the electrons as in the previous two examples. The energy balance takes a similar form as Eqs. (6.20) and (6.21). The main difference with respect to the previous examples is the treatment of the source term, which contains the net production rates. In this case, these values are obtained from the set of reactions shown on Table 6.1.

$$\dot{n}_e = r_{ion1} n_e n_{Ar} + r_{ion2} n_e n_{Ar^*} - r_{re} n_e^2 n_{Ar^+} - r_{diss} n_e n_{Ar_2^+} \tag{6.46}$$

Table 6.1 Kinetic processes for argon discharge

Process	Reaction	Rate coefficient
Elastic scattering	$Ar + e \rightarrow Ar + e$	$r_{elastic}$
Ionization 1	$Ar + e \rightarrow Ar^+ + 2e$	r_{ion1}
Ionization 2	$Ar^* + e \rightarrow Ar^+ + 2e$	r_{ion2}
Three-body recombination	$Ar^+ + 2e \rightarrow Ar + e$	r_{re}
Molecular ion conversion 1	$Ar^+ + 2Ar \rightarrow Ar_2^+ + Ar$	r_{mol1}
Molecular ion conversion 2	$2Ar^* \rightarrow Ar_2^+ + e$	r_{mol2}
Excitation	$Ar + e \rightarrow Ar^* + e$	r_{excit}
De-excitation	$Ar^* + e \rightarrow Ar + e$	$r_{deexcit}$
Dissociate recombination	$Ar_2^+ + e \rightarrow Ar^* + Ar$	r_{diss}

Source: Farouk et al. (2006).

$$\dot{n}_{Ar^+} = r_{ion1} n_e n_{Ar} + r_{ion2} n_e n_{Ar^*} + r_{mol2} n_{Ar^*}^2 - n_{re} n_e^2 n_{Ar^+}$$
$$- r_{mol1} n_{Ar}^2 n_{Ar^+} \tag{6.47}$$

$$\dot{n}_{Ar_2^+} = r_{mol1} n_{Ar^+} n_{Ar}^2 + r_{mol2} n_{Ar^*}^2 - r_{diss} n_e n_{Ar_2^+} \tag{6.48}$$

$$\dot{n}_{Ar^*} = r_{excit} n_{Ar} n_e + r_{diss} n_e n_{Ar_2^+} - r_{deexcit} n_e n_{Ar^*} - r_{mol2} n_{Ar^*}^2 \tag{6.49}$$

$$\dot{n}_{Ar} = r_{deexcit} n_e n_{Ar^*} + r_{re} n_{Ar^+} n_e^2 + r_{diss} n_{Ar_2^+} n_e - r_{excit} n_{Ar} n_e$$
$$- r_{ion} n_e n_{Ar} - r_{mol1} n_{Ar^+} n_{Ar^+}^2 \tag{6.50}$$

Finally, the Poisson equation is given by:

$$\nabla^2 V = -\frac{e}{\epsilon_0} \left(n_{Ar^+} + n_{Ar_2^+} - n_e \right) \tag{6.51}$$

Details of simulation, including boundary and initial conditions are found in Farouk et al. (2006). It is observed that as the number of species being analyzed grows, the complexity of the solution increases significantly.

It is important to note that if the application includes negative ions, they are treated in a similar manner as shown in this example, by calculating their net reaction rate and subtracting their number density in the Poisson equation. Also, in this example, the rate coefficients shown on Table 6.1 were obtained from published works; however, these coefficients can be calculated. For instance, if we consider the first net production rate equation (Eq. (6.46)), it is seen that the right-hand side is composed of a sum of terms with the form:

$$r_{ion1} n_e n_{Ar} + \cdots \tag{6.52}$$

where the rate coefficient r_{ion1} relates the reaction due to the collision between an electron and a neutral atom Ar, as seen in the "Ionization 1" process in Table 6.1. The other terms in the sum account for the reactions of electrons with other species. The reaction rate coefficients can be calculated as:

$$\langle \sigma v \rangle = \frac{1}{n} \int_{-\infty}^{\infty} \sigma(v) v f(v) dv \tag{6.53}$$

where n is the number density, v is the particle velocity, and $f(v)$ is the VDF, which in the case of a Maxwellian distribution would take the form of Eq. (6.11). The quantity $\sigma(v)$ is the *collision cross section* which is one of the fundamental parameters in the study of plasmas and is usually a function of velocity (Roth, 1995). Available computational tools, such as BOLSIG+ allow the calculation of electron transport coefficients and collision rate coefficients from fundamental cross section data.

For applications with a large number of species and reactions, several software applications are available that solve the Boltzmann's equation and allow calculating the reaction kinetics of large number of reactions and species. Some of these tools include ZDPlasKin, EEDF, BOLOS, PumpKin, and CRANE, among others.

References

T.N. An, E. Marode, and P.C. Johnson. Monte Carlo simulation of electrons within the cathode fall of a glow discharge in helium. *Journal of Physics D: Applied Physics*, 10:2317–2328, 1977.

C.K. Birdsall and A.B. Langdon. *Plasma Physics Via Computer Simulation.* Taylor & Francis, 2004.

A. Bogaerts, C. De Bie, M. Eckert, V. Georgieva, T. Martens, E. Neyts, and S. Tinck. Modeling of the plasma chemistry and plasma-surface interactions in reactive plasmas. *Pure and Applied Chemistry*, 82(6):1283–1299, 2010.

L. Brieda. *Plasma Simulations by Example.* CRC Press, 2019. https://doi.org/10.1201/9780429439780.

C. Chazelas, J.P. Trelles, I. Choquet, and A. Vardelle. Main issues for a fully predictive plasma spray torch model and numerical considerations. *Plasma Chemistry and Plasma Processing*, 37:627–651, 2017.

D.M. Chen, K.C. Hsu, and E. Pfender. Two-temperature modeling of an arc plasma reactor. *Plasma Chemistry and Plasma Processing*, 1:295–314, 1981.

T. Farouk, B. Farouk, D. Staack, A. Gutsol, and A. Fridman. Simulation of DC atmospheric pressure argon micro glow-discharge. *Plasma Sources Science and Technology*, 15(4):676–688, 2006.

E. Gogolides and H.H. Sawin. Continuum modeling of radio-frequency glow discharges. I. Theory and results for electropositive and electronegative gases. *Journal of Applied Physics*, 72(9):3971–3987, 1992.

E.P. Hammond, K. Mahesh, and P. Moin. A numerical method to simulate radio-frequency plasma discharges. *Journal of Computational Physics*, 176:402–429, 2002.

T.E. Nitschke and D.B. Graves. A comparison of particle in cell and fluid model simulations of low-pressure radio frequency discharges. *Journal of Applied Physics*, 76(10):5646–5660, 1994.

Y.P. Raizer. *Gas Discharge Physics*. Springer, 1991.

J.R. Roth. *Industrial Plasma Engineering*, volume 1. Institute of Physics Publishing, 1995.

D.L. Scharfetter and H.K. Gummel. Large-signal analysis of a silicon read diode oscillator. *IEEE Transactions on Electron Devices*, 16(1):64–77, 1969.

J.J. Shi and M.G. Kong. Cathode fall characteristics in a DC atmospheric pressure glow discharge. *Journal of Applied Physics*, 94(9):5504–5513, 2003.

T. Tajima. *Computational Plasma Physics*. Westview Press, 2004.

J.P. Trelles, E. Pfender, and J.V.R. Heberlein. A multiscale finite element modeling of ARC dynamics in a DC plasma torch. *Plasma Chemistry and Plasma Processing*, 26:557–575, 2006.

J.P. Trelles, C. Chazelas, A. Vardelle, and J.V.R. Heberlein. ARC plasma torch modeling. *Journal of Thermal Spray Technology*, 18(5–6):728–752, 2009.

A. Venkattraman. Cathode fall model and current-voltage characteristics of field emission driven direct current microplasmas. *Physics of Plasmas*, 20(11):113505, 2013.

A.L. Ward. Effect of space charge in cold-cathode gas discharges. *Physical Review*, 112(6):1852–1857, 1958.

A.L. Ward. Calculations of cathode-fall characteristics. *Journal of Applied Physics*, 33(9):2789–2794, 1962.

7

Control of Plasma Systems

7.1 Control of Thermal Plasma Torches

Thermal plasma torches are energy-intensive devices that operate at temperatures unreachable by conventional combustion or partial oxidation processes. Plasma gasification reactors for biomass are relatively easy to control, since electric power can be varied quickly to maintain conditions of the biomass conversion process. On the other hand, the physics involved in these systems are quite complex, involving the coupling of fluid dynamics, heat and mass transfer, as well as Maxwell's equations. In addition, the physicochemical transformations of the biomass in the reactor affect the operation of the torch, with all these happening at time scales that cover a wide range of orders of magnitude.

Before an attempt is made to develop an active control scheme for such systems, sources of instabilities and methods of stabilization must be analyzed. Figure 7.1 shows a plasma arc inside a coaxial anode being cooled by the flow of water. This configuration is referred to as *wall-stabilized arc*, where the mechanism of stabilization is based on the Elenbaas–Heller equation that indicates that when the arc has an asymmetric perturbation and part of it approaches the wall, increased cooling at the edge of the arc produces a rise of temperature at the axis. The heat up of the axial temperature increases the electrical conductivity of the gas and this makes the arc displace back to the axis, stabilizing the arc (Roth, 1995). Segmenting the anode wall in several parts along its length is another effective technique for stabilizing the arc. These techniques are based on cooling the edges of the arc, however, stabilizing techniques can also be based on the addition of external flows. For instance, transpiration-stabilized arcs work on the principle of adding cooling water or gas through annular slots in the radial inward direction along the length of the arc. Another technique, referred to as *coaxial flow stabilized arc*, involves the addition of a gas flow injected

Voltage-Enhanced Processing of Biomass and Biochar, First Edition. Gerardo Diaz.
© 2022 John Wiley & Sons Ltd. Published 2022 by John Wiley & Sons Ltd.

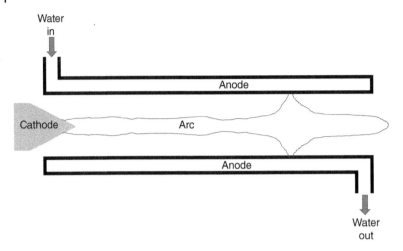

Figure 7.1 Wall-stabilized arc.

coaxially and flowing along the periphery of the arc (Fridman and Kennedy, 2004).

For the case of nontransferred arcs and plasma torches, several stabilization techniques have also been developed. For instance, in the *vortex-stabilized arc*, gas is injected tangentially to the arc diameter, forming a vortex around the arc that stabilizes it, making the wall temperature of the anode have very little effect on the operation of the torch. A more complex stabilization method involves the generation of an axial magnetic field $\mathbf{J} \times \mathbf{B}$ which rotates a wall-stabilized arc to reduce fluctuations. This method is called *magnetically stabilized rotating arcs*.

7.1.1 Dynamics

These techniques have proven effective in helping stabilize plasma arcs. However, oscillating behavior has been observed in the operation of thermal torches, which has been shown to behave chaotically. In addition, the arc may be extinguished under certain operating conditions, requiring it to be restarted. A detailed analysis of the dynamics involved in nontransferred arcs was performed by Ghorui and Das (2004), Ghorui et al. (2004). They determined that some of the oscillations observed in plasma torches are generated by the intrinsic operation of the device. Figure 7.2 shows a schematic of a nontransferred-arc plasma torch, where the arc extends from a local spot at the cathode to a point of contact at the anode, referred to as *arc root*, and where the expanding gases leave the hollow cathode as a jet.

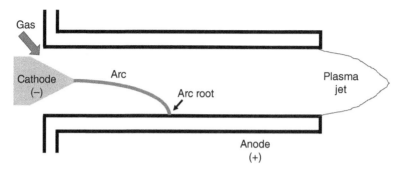

Figure 7.2 Arc root location in a thermal plasma device.

Aerodynamic drag forces the arc to the end of the exit nozzle, extending the arc and therefore increasing voltage, while the electromagnetic body force tries to keep the arc root at minimum voltage, thus trying to prevent the arc from growing. Therefore, the arc root operates in unstable equilibrium which can be easily broken by small perturbations that make the arc move to find another equilibrium root. Since plasma torches are generally operated at constant current, a fluctuation of the voltage implies that there is a fluctuation of the total power delivered to the system. Fluctuation in the power implies that the conditions for ionization also vary due to changes in electric field and collision rates. The analysis also focusses on the arc root where the high current density arc is in contact with the material of the anode (usually a metal). At the point of contact, there is heat transfer from the arc at a temperature higher than the melting point of the anode's metal. Heat is dissipated by conduction, convection, and evaporation from the anode with the vapors diffusing into the gas inside the torch, heating it up and generating an expansive force. The small contact area at the arc root for the high current density arc generates a compressive force associated with the self-magnetic field. The interaction between the compressive and expansive forces generates a highly nonlinear variation of the conditions which are affected by the operating temperatures and varying parameters of the plasma gas, such as viscosity, thermal and electrical conductivities, and heat and mass diffusivities of the material, among others.

The main governing equations for the analysis include:

Conservation of energy :

$$\frac{\partial T}{\partial t} - w\frac{\Delta T}{d} + \mathbf{v} \cdot \nabla T = \kappa \nabla^2 T \tag{7.1}$$

Conservation of metal :

$$\frac{\partial S}{\partial t} - w\frac{\Delta S}{d} + \mathbf{v} \cdot \nabla S = \kappa_S \nabla^2 S \tag{7.2}$$

Conservation of momentum :

$$\frac{\partial \mathbf{v}}{\partial t} + \mathbf{v} \cdot \nabla \mathbf{v} = -\frac{1}{\rho_0} \nabla p + \nu \nabla^2 \mathbf{v}$$
$$+ \mathbf{g}(\alpha T - \beta S) + \frac{1}{\rho_0} \mathbf{J} \times \mathbf{B} \qquad (7.3)$$

Conservation of mass :

$$\nabla \cdot \mathbf{v} = 0 \qquad (7.4)$$

Maxwell's equations :

$$\mathbf{J} = \frac{1}{\mu_0}(\nabla \times \mathbf{B}) \qquad (7.5)$$

$$\nabla \times \mathbf{E} = \frac{\partial \mathbf{B}}{\partial t} \qquad (7.6)$$

$$\nabla \cdot \mathbf{B} = 0 \qquad (7.7)$$

$$\nabla \cdot \mathbf{E} = \frac{\rho C}{\epsilon} \qquad (7.8)$$

Current density :

$$\mathbf{J} = \sigma_C(\mathbf{E} + \mathbf{v} \times \mathbf{B}) \qquad (7.9)$$

Equation of state :

$$\Delta \rho = \rho_0(-\alpha T + \beta S) \qquad (7.10)$$

where κ is the thermal diffusivity and κ_S is the diffusivity of evaporated material. After the set of equations is nondimensionalized, the nonlinear evolution of the instability amplitude can be shown to be given by Eq. (7.11):

$$\dddot{F} + \Omega_2 \ddot{F} + \Omega_1 \dot{F} + \Omega_0 F = \pm F^3 \qquad (7.11)$$

where coefficients Ω_i, for $i = 0$–2, are given by the thermophysical properties of the generated plasma, the temperature, and the magnitude of the arc current (Ghorui and Das, 2004). Analysis shows that a steady solution is obtained for $-\Omega_0 < \Omega_1/2$, while the condition $-\Omega_0 > \Omega_1/2$ gives a wide variety of oscillatory behaviors that can be obtained mainly by varying the value of Ω_0. For low values of $-\Omega_0$, the systems exhibit no oscillations. As $-\Omega_0$ is increased, oscillations become more and more complicated through a process of bifurcations, and for large values of this parameter, chaotic behavior is obtained through a period doubling route with no significant change in the attractor structure. As $-\Omega_0$ is set to a value of 144 a discontinuity occurs and the system jumps to an unbounded attractor. This catastrophic crisis can be associated with extinction of the arc.

7.1.2 Control

As opposed to conventional thermochemical processes, plasma gasification systems involve physical and chemical phenomena that occur in a very large range of time scales. Temperature fluctuations inside the biomass conversion reactor vary at time scales of the order of seconds to minutes, while pressure changes occur in hundredths to tenths of a second. On the other hand, the source of heat comes from the arc at the plasma torch, where voltages and currents can be varied at time scales of the order of milliseconds or less, and ionization and recombination reactions occur at even shorter time scales. In addition, Section 7.1.1 has established that under certain operating conditions, fluctuations at the arc root can occur and the system can display chaotic behavior. Therefore, in order to control the overall biomass conversion process, it is essential to first control the operation of the plasma torch.

Control systems are usually divided into open and closed loop. Open loop systems tend to establish a relation between input parameters and output effects. This type of control is of very limited use for highly varying and unstable systems. Therefore, closed loop approaches, where feedback control is implemented based on measurements of output variables and states to generate a corrective action, are usually more commonly found in the literature for control of plasma torches (Fincke et al., 2001, Jain et al., 2016), where proportional–integral–differential (PID)-based controllers continue to be used in thermal plasma steam reforming systems (Tsai et al., 2007, Diaz, 2016) and for real-time control of plasma cutting processing (Soylak, 2016). More recently, artificial intelligence (AI)-based control schemes have been developed to stabilize the chaotic dynamics observed in rod-type plasma torches (Kim, 2019, Pai et al., 2010, Salahshour et al., 2019-9). Since in industrial systems the information of the states of the plant is not always available, other researchers have introduced reduced-order nonlinear observers in the stabilization of these chaotic systems (Malekzadeh and Noei, 2021).

7.1.2.1 Development of Fractional Order Controller for Chaotic Behavior

The development of controllers and dynamical models of the plants has traditionally been based on integer-order derivatives. However, in certain

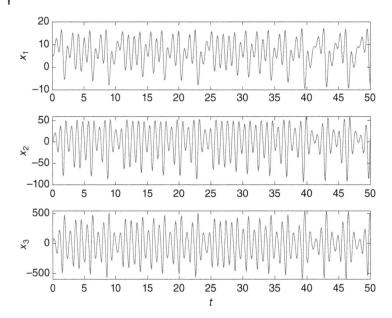

Figure 7.3 Time series for rod-type plasma torch.

applications, fractional order dynamical models and controllers have shown advantages with respect to conventional approaches. Here, a fractional order controller is developed to stabilize the chaotic behavior of a rod-type plasma torch.

The dynamical model of such systems is given by Malekzadeh and Noei (2021):

$$
\begin{aligned}
\dot{x}_1 &= x_2 \\
\dot{x}_2 &= x_3 \\
\dot{x}_3 &= -ax_1 - bx_2 - cx_3 + dx_1^3 + u
\end{aligned}
\tag{7.12}
$$

where u is a single-variable control input and x_i, $i = 1,2,3$ are states of the system. For an uncontrolled system where $u = 0$, Figure 7.3 shows the chaotic behavior of the system as a time series, where Figure 7.4 shows the attractor formed for parameter values: $a = -130$, $b = 50$, $c = 1$, and $d = -1$ and for an initial condition $x(0) = [5, -2, 3]^T$.

The fixed points of this system are $\mathcal{O}_1 = (0,0,0)$, $\mathcal{O}_2 = (-\sqrt{a/d}, 0, 0)$, $\mathcal{O}_3 = (+\sqrt{a/d}, 0, 0)$. Following the procedure described in Tavazoei et al. (2009) and Diaz and Coimbra (2010), a fractional order integrator is developed to stabilize the single-input chaotic system for any of the nontrivial

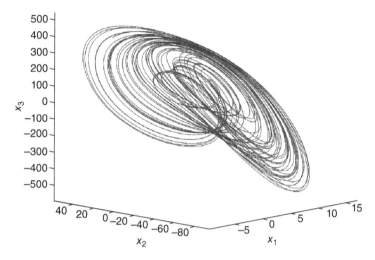

Figure 7.4 Attractor for rod-type plasma torch.

fixed points. The Jacobian matrix evaluated at either $\mathcal{O}_2$ or $\mathcal{O}_3$ is given by:

$$A = \begin{bmatrix} 0 & 1 & 0 \\ 0 & 0 & 1 \\ 2a & -b & -c \end{bmatrix} \tag{7.13}$$

where the characteristic polynomial of the Jacobian matrix is given by

$$s^3 - cs^2 - bs + 2a = 0 \tag{7.14}$$

The linearized system can be written as:

$$\frac{d\eta}{dt} = A\eta + Bu \tag{7.15}$$

where $\eta = x - x^*$, x^* is a fixed point, and

$$B = \begin{bmatrix} 0 \\ 0 \\ 1 \end{bmatrix} \tag{7.16}$$

The linear system (7.15) can be stabilized by using a state fractional integrator control input given by:

$$u(t) = -J^q(\mu\eta_1 + v\eta_3) \tag{7.17}$$

where q is a rational number between 0 and 1, J^q is the qth order fractional integral operator, and μ and v are two positive constants, which can be chosen following Tavazoei et al. (2009) as:

$$0 < \mu < \frac{cb^{1+q/2}}{\cos(q\pi/2)} \quad \text{and} \quad v > \frac{(2a)b^{-1+q/2}}{\cos(q\pi/2)} \tag{7.18}$$

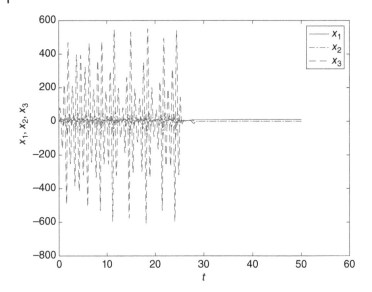

Figure 7.5 Fractional order control for the chaotic rod-type plasma torch system.

to make the system asymptotically stable. Choosing an arbitrary value of $q = -0.2$ with the variable order operator given by the expression (Diaz and Coimbra, 2009, 2010, Coimbra, 2003):

$$\mathcal{D}^{q(t)}x(t) = \frac{1}{\Gamma(1 - q(t))} \int_{0+}^{t} (t - \sigma)^{-q(t)} \mathcal{D}^1 x(\sigma) d\sigma + \frac{(x(0^+) - x(0^-))t^{-q(t)}}{\Gamma(1 - q(t))}$$

(7.19)

which is valid for $q(x(t)) < 1$. It is noted that positive values of $0 < q < 1$ produce a fractional order derivative while negative values produce fractional integration. Using condition (7.18) to select $\mu = 1.3994$ and $v = 8.8938$, the chaotic system can be stabilized as shown in Figure 7.5 where the controller has been turned on at a dimensionless time $t = 25$.

7.2 Control of Nonthermal Plasma Discharges

Nonthermal plasmas have traditionally been used under vacuum conditions where a gas such as argon, helium, or other, is injected at a low flow rate for performing processes that induce plasma chemistry or modify the surface characteristics of a material (Sugawara et al., 1998, Meichsner et al., 2013, Lieberman and Lichtenberg, 2005). However, for energy-related applications, atmospheric-pressure conditions are required for the purpose of reducing reactor cost and for allowing continuous operation to be achieved as opposed to batch processing. The literature on atmospheric pressure nonthermal plasmas has seen a tremendous growth in the past decades (Wolf, 2012, Chu and Lu, 2013). Nonthermal plasmas can be generated from a variety of sources, with nanosecond pulsed plasma being a source of discharges that has been used for pollution control for NO_x and SO_x removal, volatile organic compounds (VOCs) decomposition, and odor control. Pulsed plasma has been analyzed numerically (Okubo, 2015) as well as with experiments at waste disposal pilot-scale plants (Yoshida et al., 2009), among other research works. Results have shown reduction in NO_x concentration of more than 90%. However, nanosecond pulse plasma equipment is relatively expensive, which has motivated researchers to use less costly sources of nonthermal plasma such as dielectric barrier discharge (DBD) and radio frequency (RF) power supplies. In many applications, nonthermal plasma is used as the control action, especially for flow control (Neretti, 2016, Wang et al., 2013, Adamiak, 2020), where plasma is used at control surfaces to modify drag and flow characteristics. Nonetheless, in energy related applications, nonequilibrium plasma is usually used as a tool for pollution control to destroy or decompose VOCs, NO_x, and other contaminants, or for ozone generation. Saleem et al. (2019) analyzed the decomposition of model tars from biomass gasification using DBD. The plasma generator was controlled using a Variac AC transformer to modify the input voltage. The power supply could deliver from 5 to 40 W at a frequency of 20 kHz. In general, plasma parameters as well as other physical inputs to the reactor need to be controlled in real time. Depending on the power supply utilized, plasma related quantities such as input voltage, frequency, or current can be manually modified or can be varied by a feedback control scheme. In addition, mass flow reactor temperature, and pressure can have their separate control loops which can be coupled or completely decoupled.

7.2.1 Plasma Diagnostics

One of the requirements for the implementation of feedback control is to be able to measure a desired output that can be compared with the reference value to make adjustments to the control action that reduces the difference between them. Some controllers, e.g. proportional, do not guarantee zero offset, and in many cases, an integral action is required. Whether model based control, AI-assisted, PID, single or multivariable, or any other type of control is used to stabilize plasma, adequate diagnostic techniques that provide information about the states of the system are needed to generate feedback signals to the controller. As opposed to thermal arcs where it is very hard to use invasive measurement techniques, nonthermal plasmas allow the use of both invasive and noninvasive techniques for characterization purposes. One of the traditional invasive techniques is the *Langmuir probe* which can provide information about plasma density, electron temperature, and plasma potential. Other available diagnostic tools are *Mass Spectrometer and Energy Analyzer* to identify positive and negative ions, as well as neutrals and radicals, *Optical Pyrometer* to measure high temperatures, *Microwave interferometer* for plasma density measurements, and *UV radiometer* to determine ultraviolet light spectrum, among others. For a detailed description of plasma diagnostics, the reader is referred to Hutchinson (2005) and Keidar and Beilis (2013).

In addition, to the measurements of plasma quantities, gas analyzers for gas composition, gas chromatography, gas spectrometers, mass spectrometers, etc. could also be used to measure the product gas composition before and after the plasma reactor. These measurements can be utilized for feedback control.

7.2.2 AI-Based Control

The traditional feedback control system is depicted in Figure 7.6 where the measurement y_m is subtracted from reference r to form the error e that is provided to the *controller*. The controller generates a control action u that is the input to the *plant* that produces the output y. This output is measured by sensors to generate y_m. The control scheme tries to reduce the difference between reference (r) and y_m.

This configuration is general and the same concept can be applied to systems with single or multiple inputs and outputs.

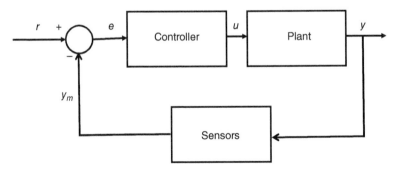

Figure 7.6 Conventional feedback control configuration.

The reduction in the cost of sensors and the consequent introduction of the *Big Data* concept, which relates to the existence of volumes of data so large and complex that they become impossible to process with traditional analysis methods, have accelerated the development of AI techniques such as deep neural networks, genetic algorithms, fuzzy logic, and the different variations of autoregressive moving average models (ARMA, ARMAX, ARIMA, etc.) that can process large amounts of data. The development of these models has contributed to the introduction of control techniques that take advantages of these models that can be retrained to adapt to changing conditions. *Internal model control* corresponds to a type of feedback control that adapts very well to the use of AI models as shown in Figure 7.7, where a model of the plant is used to obtain an error that is fed back to the controller which generates the control action. In general, disturbances can be present which might require real-time modification of the AI model.

For these models, the more data, the better approximation of the AI model to the real plant, although aspects such as overfitting and data generalization need to be always considered. The controller shown in Figure 7.7 can be applied to a variety of thermal systems (Diaz et al., 2001), and in particular to thermal and nonthermal plasma reactors (Mesbah and Graves, 2019), and can be of the conventional type, such as PID, H_2, H_∞, fractional, etc. or it can be another AI model trained as a controller.

In the case of *artificial neural networks* (ANNs), static models involve an array of input parameters relevant to the plant, and the ANN provides a desired output value, which is a steady-state prediction based on the inputs provided with no consideration of time. For the case of training the ANN for dynamic behavior, one method consists on providing the time explicitly

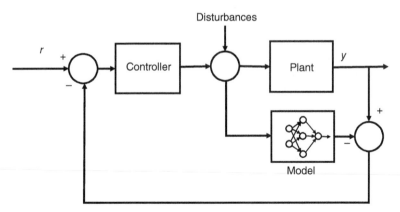

Figure 7.7 Internal model control configuration. Source: Based on Diaz et al. (2001).

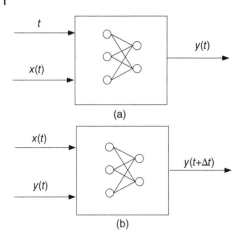

Figure 7.8 ANN training methods for dynamic simulations. (a) Explicit time. (b) Implicit time.

as an input t and the values of the states $x(t)$ at that time, while the ANN is trained to predict the plant's output $y(t)$, as shown in Figure 7.8a. This methodology performs well when the tests involve a specific range of time. However, in most cases, the model is required to run continuously for a long time, so in these cases, the time is included implicitly, where the values of the state $x(t)$ and current output $y(t)$ are provided as inputs while the ANN is trained to predict the output at the next time step $y(t + \Delta t)$, as shown in Figure 7.8b. It is observed that this configuration can be used to provide input information at different previous time steps such as $x(t)$, $x(t - \Delta t), x(t - 2\Delta t), \dots, y(t), y(t - \Delta t), y(t - 2\Delta t), \dots$ in order to improve the accuracy of the prediction of $y(t + \Delta t)$.

7.2.2.1 ANN Model of a Thermal Plasma Torch

The dynamic behavior of a thermal plasma discharge in contact with biomass, shown in Figure 4.4, is revisited to develop an AI model based on artificial neural networks. An ANN is trained with the format shown in Figure 7.8b, where an input layer contains three inputs, the first one being the information of the status of the plasma discharge ($i = 1$ for "on" and $i = 0$ for "off"), while the second and third inputs correspond to the temperature of the discharge at the current and previous time step ($T(t), T(t - \Delta t)$). The single output of the ANN corresponds to the discharge temperature at the next time step ($T(t + \Delta t)$). In this way,

$$T(t + \Delta t) = \mathcal{F}\,(i, T(t), T(t - \Delta t)) \tag{7.20}$$

where $\mathcal{F}$ is a function represented by the ANN. The ANN is trained using the Levenberg–Marquardt backpropagation algorithm, where a fraction of the test data is used for training and the rest for testing. Utilizing a single

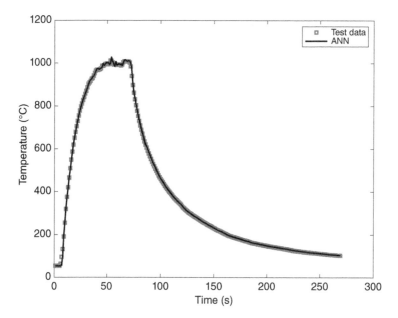

Figure 7.9 Dynamic model of a thermal plasma torch based on ANNs.

hidden layer of 30 nodes, the ANN is tasked to predict the temperature in the next time step. The plasma torch is kept turned on ($i = 1$) until ($t = 73$ seconds) and then it is turned off for the rest of the time ($i = 0$). Figure 7.9 shows the comparison of the test data with the results obtained with the ANN. The ANN has been able to learn the dynamic behavior the discharge temperature for the data obtained with the thermocouple that was positioned at the shortest distance from the nozzle. The model can be utilized to learn the behavior of the temperature at other locations.

This type of modeling tools are exceptionally capable for modeling complex physical phenomena where abundance of data is available. The model can be later on used in an internal model control scheme (IMC) to drive the discharge to a desired operating condition.

References

K. Adamiak. A simplified model for simulating flow stabilization behind a cylinder using dielectric barrier discharge. *IEEE Transactions on Plasma Science*, 48(7):2464–2474, 2020. https://doi.org/10.1109/TPS.2020.2998061.

P.K. Chu and X. Lu. *Low Temperature Plasma Technology*. CRC Press, 2013.

C.F.M. Coimbra. Mechanics with variable-order differential operators. *Annalen der Physik*, 12(11–12):692–703, 2003.

G. Diaz. Plasma steam reforming. In *CRC Encyclopedia of Plasma Technology*. Ed. J.L. Shohet. Taylor and Francis, 2016.

G. Diaz and C.F.M. Coimbra. Nonlinear dynamics and control of a variable order oscillator with application to the van der Pol equation. *Nonlinear Dynamics*, 56(1):145–157, 2009.

G. Diaz and C.F.M. Coimbra. Dynamics and control of nonlinear variable order oscillators. In *Nonlinear Dynamics*. Ed. T. Evans, pages 129–144. INTECH, 2010.

G. Diaz, M. Sen, K.T. Yang, and R.L. McClain. Dynamic prediction and control of heat exchangers using artificial neural networks. *International Journal of Heat and Mass Transfer*, 44(9):1671–1679, 2001. ISSN 0017-9310. https://doi.org/10.1016/S0017-9310(00)00228-3.

J.R. Fincke, W.D. Swank, R.L. Bewley, D.C. Haggard, M. Gevelber, and D. Wroblewski. Diagnostics and control in the thermal spray process. *Surface and Coatings Technology*, 146–147:537–543, 2001. ISSN 0257-8972. https://doi.org/10.1016/S0257-8972(01)01432-3. Proceedings of the 28th International Conference on Metallurgic Coatings and Thin Films.

A. Fridman and L.A. Kennedy. *Plasma Physcis and Engineering*. Taylor & Francis Books, 2004.

S. Ghorui and A.K. Das. Theory of dynamic behavior in atmospheric pressure arc plasma devices: Part I: Theory and system behavior. *IEEE Transactions on Plasma Science*, 32(1):296–307, 2004.

S. Ghorui, S.N. Sahasrabudhe, P.S.S. Murty, and A.K. Das. Theory of dynamic behavior in atmospheric pressure arc plasma devices: Part-II: Validation of theory with experimental data. *IEEE Transactions on Plasma Science*, 32(1):308–315, 2004.

I.H. Hutchinson. *Principles of Plasma Diagnostics*. Cambridge University Press, 2nd ed., 2005.

V. Jain, S.K. Nema, and V. Agarwal. Design and simulation of feedback system to generate plasma arc in current source mode. In *2016 IEEE 7th Power India International Conference (PIICON)*, pages 1–5. IEEE, 2016.

M. Keidar and I.I. Beilis. Chapter 2 - Plasma diagnostics. In *Plasma Engineering*. Ed. M. Keidar and I.I. Beilis, pages 83–101. Academic Press, Boston, MA, 2013. ISBN 978-0-12-385977-8. https://doi.org/10.1016/B978-0-12-385977-8.00002-0.

K.S. Kim. Control-oriented dynamic model of an inductively coupled plasma torch by artificial intelligence methodology. *Plasma Physics and Controlled Fusion*, 61(4):044002, 2019. https://doi.org/10.1088/1361-6587/aaffb4.

M.A. Lieberman and A.J. Lichtenberg. *Principles of Plasma Discharges and Materials Processing*. Wiley, 2nd ed., 2005.

M. Malekzadeh and A.R. Noei. Nonlinear observer design based on immersion and invariance method: an insight to chaotic systems. *International Journal of Dynamics and Control*, 9(438–447), 2021. https://doi.org/10.1007/s40435-020-00670-7.

J. Meichsner, M. Schmidt, R. Schneider, and H.-E. Wagner. *Nonthermal Plasma Chemistry and Physics*. CRC Press, 2013.

A. Mesbah and D.B. Graves. Machine learning for modeling, diagnostics, and control of non-equilibrium plasmas. *Journal of Physics D: Applied Physics*, 52(30LT02):1–9, 2019.

G. Neretti. Active flow control by using plasma actuators. In *Recent Progress in Some Aircraft Technologies*. Ed. R.K. Agarwal. IntechOpen, 2016.

M. Okubo. Evolution of streamer groups in nonthermal plasma. *Physics of Plasma*, 22(123515):1–6, 2015.

N.-S. Pai, H.-T. Yau, and C.-L. Kuo. Fuzzy logic combining controller design for chaos control of a rod-type plasma torch system. *Expert Systems with Applications*, 37(12):8278–8283, 2010. ISSN 0957-4174. https://doi.org/10.1016/j.eswa.2010.05.057.

J.R. Roth. *Industrial Plasma Engineering*, vol. 1. Institute of Physics Publishing, 1995.

E. Salahshour, M. Malekzadeh, R. Gholipour, and S. Khorashadizadeh. Designing multi-layer quantum neural network controller for chaos control of rod-type plasma torch system using improved particle swarm optimization. *Evolving Systems*,10(3), 2019. ISSN 1868-6478.

F. Saleem, K. Zhang, and A. Harvey. Plasma-assisted decomposition of a biomass gasification tar analogue into lower hydrocarbons in a synthetic product gas using a dielectric barrier discharge reactor. *Fuel*, 235:1412–1419, 2019.

M. Soylak. A novel real-time control system for plasma cutting robot with xpc target. *Advances in Mechanical Engineering*, 8(4):1–12, 2016.

M. Sugawara, B.L. Stansfield, S. Handa, K. Fujita, S. Watanabe, and T. Tsukamoto. *Plasma Etching: Fundamentals and Applications, Series on Semiconductor Science and Technology, 7*. Oxford University Press, 1998.

M.S. Tavazoei, M. Haeri, S. Bolouki, and M. Siami. A novel real-time control system for plasma cutting robot with XPC target. *Using Fractional-Order Integrator to Control Chaos in Single-Input Chaotic Systems*, 55:179–190, 2009.

H.-L. Tsai, C.-S. Wang, and P.M. Duc. Control design of ethanol steam reforming in thermal plasma reformer. In *16th IEEE International Conference on Control Applications*, pages 706–711. IEEE, 2007.

J.-J. Wang, K.-S. Choi, L.-H. Feng, T.N. Jukes, and R.D. Whalley. Recent developments in DBD plasma flow control. *Progress in Aerospace Sciences*, 62:52–78, 2013. ISSN 0376-0421. https://doi.org/10.1016/j.paerosci.2013.05.003.

R.A. Wolf. *Atmospheric Pressure Plasma for Surface Modification*. Wiley, 2012.

K. Yoshida, T. Yamamoto, T. Kuroki, and M. Okubo. Pilot-scale experiment for simultaneous dioxin and NO_x removal from garbage incinerator emissions using the pulse corona induced plasma chemical process. *Plasma Chemistry and Plasma Processing*, 29:373–386, 2009.

Index

Voltage-Enhanced Processing of Biomass and Biochar, First Edition. Gerardo Diaz.
© 2022 John Wiley & Sons Ltd. Published 2022 by John Wiley & Sons Ltd.